ISBN 978-1-334-29554-6
PIBN 10686476

For support please visit www.forgottenbooks.com

1 MONTH OF
FREE
READING

at

www.ForgottenBooks.com

By purchasing this book you are eligible for one month membership to ForgottenBooks.com, giving you unlimited access to our entire collection of over 1,000,000 titles via our web site and mobile apps.

To claim your free month visit:

www.forgottenbooks.com/free686476

English
Français
Deutsche
Italiano
Español
Português

www.forgottenbooks.com

Mythology Photography **Fiction**
Fishing Christianity **Art** Cooking
Essays Buddhism Freemasonry
Medicine **Biology** Music **Ancient
Egypt** Evolution Carpentry Physics
Dance Geology **Mathematics** Fitness
Shakespeare **Folklore** Yoga Marketing
Confidence Immortality Biographies
Poetry **Psychology** Witchcraft
Electronics Chemistry History **Law**
Accounting **Philosophy** Anthropology
Alchemy Drama Quantum Mechanics
Atheism Sexual Health **Ancient History**
Entrepreneurship Languages Sport
Paleontology Needlework Islam
Metaphysics Investment Archaeology
Parenting Statistics Criminology
Motivational

25319

APPLIED SCIENCE

INCORPORATED WITH

TRANSACTIONS OF THE UNIVERSITY OF TORONTO ENGINEERING SOCIETY

Vol. IV - New Series

November '10 to April '11

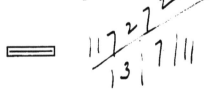

PUBLISHED BY

THE UNIVERSITY OF TORONTO ENGINEERING SOCIETY

ENGINEERING BUILDING
UNIVERSITY OF TORONTO, CANADA

TA.
|
S5
no.23

INDEX VOL. IV.

Applied Science

INCORPORATED WITH

TRANSACTIONS OF THE UNIVERSITY OF TORONTO ENGINEERING SOCIETY

Old Series Vol. 23   NOVEMBER, 1910   New Series Vol. IV. No. 1

THE SCOPE OF ENGINEERING IN CANADA.

R. W. LEONARD, C.E.

Mr. Chairman and Gentlemen,—I esteem it an honor to have been requested by the President of your Engineering Society to address you on some engineering subject of my own selection, and in a rash moment I consented, not realizing at the time that your Society embraced men of wide experience in the many fields of engineering since graduating from one of the foremost technical schools in America.

Since realizing the gravity of the situation and balancing it up with my natural aversion for hard work, I have decided to give a very general short address on engineering in which the points dealt with may be very stale and uninteresting to the seniors, but may be of some value to the younger members who are the only ones I can hope to influence, as I am not long out of the ranks of juniors myself, having had very little experience in some essential matters—particularly, as you will observe, in making addresses in public.

In this connection I may express my admiration of the teaching of the art of public speaking which is now given generally in some of our colleges. In my day at college, this was neglected and the result is that I have several times envied the assurance and ability displayed by youngsters of fourteen years in public speaking. This is an important matter for an educated man in any walk of life that he may be able in public to express his thoughts logically, clearly, and concisely, and with the confidence that comes of practice.

An engineer, according to the dictionary, is one who uses or has to do with the construction of engines or machines: or of works in which machinery is extensively employed. He is also sometimes defined as one who utilizes the forces of nature for the benefit of man. Someone whose energies were mostly directed toward railway location, construction and maintenance, has defined an engineer as "a man who can make a dollar do the most work." This last is a definition not to be despised as the financial result is one of the most important measures of the

---

* Read before the Engineering Society, Oct. 5th, 1910.

success of all engineering problems, and any man who cannot carry out his work satisfactorily in this regard will not be considered by his employer as a successful engineer. On the other hand, where a reasonable expenditure will place the success of a work beyond peradventure, the expense must not be spared.

Advancing civilization has so complicated man's requirements and multiplied the field of endeavor of the man whose profession of old was confined in the army to the handling of artillery, and in civil life to the construction of roads, bridges, buildings, and steam engines, that perhaps a good definition to-day of an engineer would be "a man who does things."

By the way, the Ancient Romans designated the chief of the highest members of their great colleges of priests, Pontifex Maximus, or the Chief Bridge Builder, from which we get the word Pontiff, all of which shows the early connection between the church and the civil engineer, and is only another proof of its antiquity, respectability, and honor of the profession, and explains the affinity of engineers to the church which is so conspicuous to this day.

All this raises the question if the French name "Ingineur," spelt with an I, and probably related to our word "Ingenuity," would not be a more appropriate designation for our profession. This is a suggestion which I leave to some of you who know something of philology.

Reverting to the financial definition of the term engineer, and assuming its correctness in many branches of the profession, then, an engineer's education is not complete until the engineer has had experience in executing the work as well as in the planning of it, and the measurement of it as done. In my opinion therefore, a young civil engineer who intends to specialize in public works, such as railway, canal, or dock work, should take one contract for the purpose of learning the "practical" part of the work—what the dollar is worth in labor and material, and how to use it to best advantage. He may lose money, but he will gain more valuable experience. My first appreciation of this practical knowledge was acquired through observing a successful contractor timing with his watch the trips the teams were making with the wheel scrapers, and figuring the cost of moving the earth per cubic yard therefrom.

Now, as my remarks apply to the juniors of the profession, I will venture upon some advice based on my limited experience, notwithstanding my familiarity with the saying that "advice is worthless, as the wise man does not need it, and the fool won't take it." One of the first things necessary to success in life is a full appreciation of the value of "law, order, duty and restraint, obedience, discipline," as Kipling puts it, in other words, executive ability. These are not taught or practised in all colleges, and more's the pity for the chances of success in after-life for the students. These can best be acquired through military training which is well worth your while, for these reasons alone. What a common experience it is to find a well-educated technical man absolutely unqualified to direct the operations of half a dozen men! Cultivate a spirit of absolute loyalty to your superiors, speaking of them and to them respectfully on all occasions with deference to their opinions. Require the same deference from your subordinates while on duty and this is best taught by your own example. Be equally loyal to your subordinates, paying attention to their suggestions and recommendations, and giving them due credit when you find their advice worthy of adoption. This brings out their hearty co-operation and insures their interest in the work. The advantage of such co-operation is apparent if one thinks of how little he alone can do and how much he is dependent upon the detail work of his assistants.

When given an opportunity to reorganize and carry on a work started by another, don't make radical changes too quickly. Better continue in the old methods for a little while, even if apparently faulty, until you have ample time to fully grasp the situation, and then make changes one at a time, using the men and materials on the work so far as possible.

Learn to judge men and their ability from personal observation rather than from written references, and if an assistant is not a success in the work you have set him to, try him in some other department, if he be a desirable character. By so doing you will get a staff about you whose capacities are known and on whom you can rely. This is on the old principle that " 'Tis better to live with the devils you know than to go to the devils you don't know."

Of course, no one man can get a thorough education in college in all the branches of the profession, including mechanics, mining, metallurgy, chemistry, electricity, hydraulics and sanitary engineering—as it would take a life-time.

There is a general ground work common to all, however, that can be acquired at college, and the actual education in any one branch must be obtained by the student after leaving college through experience and close study. This remark is intended to chasten any youngster who has just got his degree and who might therefore consider himself fully qualified to teach his maternal ancestor the noble art of sucking eggs. Please do not infer from this that all the older members of the profession are "grannies," as I am too nearly graduating into their ranks to make such a suggestion.

Though that tendency of some graduates is not to be commended, engineers need to have much self-confidence, and be prepared in this young country to take hold of any work in any branch of the profession that may present itself, and must have the courage to take the responsibility of that work and to carry it through to success.

A few years ago in Canada, civil engineering meant railway or canal surveys and construction, and some water supply and sewage works, and the profession consisted of a very small but select company of men, who set the pattern of loyalty to employers, of industry and integrity to the latter generation that the civil engineers in Canada to-day are proud to acknowledge and to emulate.

I remember being one of only three candidates for matriculation in the faculty of Applied Science at McGill, and I dropped out for want of necessary funds to take the course. Now the students in Applied Science in Canadian colleges are numbered by the hundreds and the Can. Soc. C.E. enrolls about 1,400 corporate members, and one of our ablest engineers estimates that "it requires the expenditure of $150,000,000.00 per annum on engineering works to keep us all employed and give us a moderate remuneration." . . . . . "And we cannot expect the long continuance of the expenditure of capital for engineering works which is required to keep us all going."

If the saying attributed to one of our foremost statesmen that "The 20th century belongs to Canada," and I believe it does in the sense of a vast increase in population, wealth and importance, then surely this increase is to be effected through the agency of "the men who do things" or the engineers, and in my mind, there can be no overproduction of engineers provided they be well equipped by a proper training along the broad lines of engineering suggested.

We must bear in mind that the opening up and developing of a new country with its consequent requirements in public works and industries of all sorts is not effected by the lawyers and the physicians (important as these professions may be), and we must see to it that our educational institutions are kept alive to the requirements of modern civilization, the necessary education for which is no longer crystallized in ancient books written in the dead languages.

To my mind this means that many of you must be prepared to take up mining, metallurgical, electro-chemical, and other industries which are best managed by the educated engineer, and that there is ample work for an immense number of men educated along the lines indicated in Canada. Consider our advance in mining alone. The mineral production of Canada in 1886 was $10,221,255; 1896, $22,474,256; 1906, $79,057,308; 1909, $90,415,763, or $12 per capita, as against $2.23 per capita in 1886—and we have just begun.

Northern Ontario and Quebec have scarcely been glanced at as yet by the prospectors, yet you are all familiar with the fact that Ontario produces the bulk of the world's nickel and about 12 per cent. of the world's silver and much of its corundum and mica. Quebec produces most of the world's asbestos.

Our prairies and British Columbia contain the coal that is

essential for the existence of the population flowing in there, and for the North Western United States.

We have many hundreds of miles of the same rocky mountains that have produced all the fabulous wealth in silver, gold and copper, in South America, Mexico, and the Western United States. In Canada the development of these undoubted mineral resources is comparatively slow, owing to climatic conditions in winter and the general distribution of forest growth owing to the moister climate—meaning so much greater things for the ultimate welfare of the country.

Here we see forest growth and forestry entering into the domain of engineering and raising the question of how nearly allied is the science of forestry and the conservation of other natural resources to our profession. Here are a few figures to measure the question by: In 1887 I built a large coal-loading wharf in Nova Scotia, the hemlock timber for which cost $4.50 per 1,000 feet B.M., sawn in bridge sizes and loaded on cars. It had been cut down for the tan bark, a part of what would otherwise have rotted in the woods was utilized for the wharf. The present price of such timber there would probably be $20 per 1,000. About 1891, in Parry Sound District, white pine hewn to railway bridge sizes was delivered at the site of the structures for $7.50 per 1,000 feet B.M. During the past two years nearly all the timber of the same sizes required for the mining buildings in Cobalt has been brought from British Columbia at a cost of about $27 per 1,000.

The "inexhaustable" pine forests of Canada are of the past as our other forests will shortly be if we do not take extraordinary precautions towards conserving and replanting them. A recent fishing trip in the northern portion of Quebec impressed me with the importance of the work being done in this respect by our professional ancestors, the beavers. They have been protected there for some years, and have increased until they are very numerous and have flooded every old-time dry beaver meadow and raised the level of every pond and small lake from two to four feet. In the aggregate this must have an important effect in checking forest fires and conserving and regulating the flow of water in the rivers, and benefiting the conditions desirable for the operation of water powers.

Coincident with the increased cost of timber we find a decrease in the cost of Portland cement of much superior quality which is rapidly taking the place of timber in many structures.

When reinforced with steel we get a most important new building material, the proper use of which is an art almost worthy of forming a separate branch of Civil Engineering.

Concrete is beginning to be used in ship-building. Barges of reinforced concrete are said to be in use in Europe, and it is said that one is being designed to be built in Ontario. It would appear that the design must be very simple, or the cost for timber forms will be excessive. The economical use of lumber for forms for concrete structures and the design of structures to this end, forms a subject well worth much more attention than has generally been given it to date.

To return to this very important subject of transportation, which a few years ago engaged almost solely the engineering skill in Canada.

Beginning about a century ago, the rapids of the St. Lawrence were canalized for small boats, and a similar canal was built at the Sault by the Hudson Bay Company. Then followed continuous enlargements, the construction of the Welland Canal begun early last century by private enterprise and afterwards enlarged by the Dominion Government, until Canada has transportation facilities for ships of over 2,000 tons carrying capacity through the greatest canal system in the world right from the Atlantic to the head of Lake Superior equally free to ships of all nations.

During the past few years the Dominion Government has had surveys made of the proposed Ottawa ship canal to take ocean ships from Montreal up the Ottawa and Mattawa Rivers to North Bay, and down the French River to Lake Huron, at an estimated cost of over $100,000,000. It would appear to me that this should be the next large national public work to be undertaken so soon as the country has the National Transcontinental Railway completed and satisfactorily financed. Its construction will be warranted as a transportation route, to say nothing of the value of the water powers to be developed thereby, and its value for defence purposes. As a comparison, consider the cost to the State of New York of $150,000,000 for the enlargement of the Erie Canal to carry barges of 1,000 tons only from Buffalo to Troy.

Our western country is greater and of more value, I believe, than that of the United States.

In railway work the Grand Trunk was the pioneer and served the southerly portions of Ontario and Quebec exclusively for many years. Then followed the Intercolonial, forming a bond of union with the Maritime Provinces, thus making possible the consolidation of the Easterly Provinces within one Dominion.

One of the terms of the Confederation of British Columbia with the Dominion was the early construction of the Canadian Pacific Railway, which had been started as a government measure in the seventies. The Canadian Pacific Railway represented by a syndicate of leading Canadians, undertook the work early in the eighties, and in a remarkably short space of time, completed the laying of the rails to the Pacific in 1885, at a time when many of the people of this country declared it would never earn enough money to pay for grease for the axles. I was on the construction of that road north of Lake Superior when the "syndicate" had exhausted its means, and for about six months we, the engineers, got no pay. Now the earnings of this, the greatest railway system in the world under one management, is close to $2,000,000 per week.

Now the National Transcontinental Railway is being constructed, duplicating our railway transportation facilities from Moncton, N.B., to the Pacific, at a distance to the north of the Grand Trunk and Canadian Pacific through Ontario and Quebec that is giving that part of Canada a width from north to south undreamed of a few years ago.

The Canadian Northern Railway is another road which in a few years will be a third transcontinental highway for Canada, the country that 25 years ago was estimated by many of our people as being unable to pay for grease for the wheels of one. At that time many of our people were leaning strongly towards "commercial union" or "annexation" with the United States as our only salvation from financial ruin.

I have been reading "The Valour of Ignorance," by Major-General Homer Lee of the U. S. Army until I am almost a believer in annexation myself, the annexation of the United States to Canada. Read the book, it will repay you.

This Canadian Northern Railway is being built by two Canadians who started in life without any special educational or other advantages than industry, any amount of pluck and an abiding faith in the future of this country. Now they are probably the greatest railway owners in the world, and bid fair in a few years to own as a partnership of two a complete transcontinental railway of several thousands of miles, to say nothing of their other enormous interests, street railway, mines, steamships, and other engineering works.

In water power development Canada ranks almost first with nearly a half-million horse-power developed or being developed in the Niagara District alone. Consider the value of this to the territory served by this power by comparing its cost at, say, an average of $25 per horse-power per year, for 24-hour service to the consumer as against the cheapest other available power, the gas engine at, say, $50 per year, or the steam engine at any price from $75 per year to $150 per horse-power per year.

In 1893 I had charge of the construction of a very large (at that time) hydro-electric plant at Niagara Falls, Ontario, for the Park & River Railway Company. We developed 2,000 horsepower and the best electrical authorities advised that it was cheaper to put in and operate a steam engine and generator at Queenston to operate cars on that grade, 12 miles from the power-house at the Falls, than to convey the current from the main power-house. It was put in and operated for some years. Now we see that same power being carried a couple of hundred miles at 110,000 volts. Have we reached the limit?

Well, the subject of engineering is so wide and Canada is so big and so full of opportunities for any one who is industrious that were I gifted as a speaker I could continue along this strain till you are all asleep.

However, I will close with another word of warning:

We are rapidly acquiring much desirable national wealth and are no longer a small number of poor people struggling for a bare existence along a northerly fringe of the United States, as we were a few years ago.

As our national wealth increases and our importance is recognized by other nations, we become more subject to the envy of those countries in Europe, Asia, or America, whose populations are sufficiently dense to desire room for expansion, and we must therefore be prepared to defend our property unless we are content to become a conquered people subject to laws and customs which are alien.

This may seem to many of you as far-fetched or exaggerated Again, I advise you read "The Valour of Ignorance," by Homer Lee.

Let me call your attention to the words of Mr. L. S. Amery, Colonial Editor of the London Times in an address at the Canadian Military Institute the other day.

"We do not," he said, "consider the work of the sanitary engineer or the physician in defending the community or the individual against disease as unproductive or unnecessary, or the work of the teacher in defending the child against ignorance, or that of the clergyman in defending his people against immorality, or of the lawyer in defending his client against imposition and injustice, as unproductive and unnecessary work. Neither should we consider the work of the nation in providing munitions of war or the work of the soldier in preparing himself by study and training to defend his home and country as unproductive work. All these are essential to national existence and to national development."

He believes that it was the duty of every able-bodied citizen to prepare himself to defend his country. "A nation that looks after the physical, intellectual and moral development of its people by means of the necessary military training will solve all of its other problems. Its industrial and commercial efficiency will be enhanced because of the patriotism and capabilities of its trained manhood. Perhaps a few people say that Canada has no need for a defence force, that Great Britain will defend this country, or this nation can lean on the United States; surely Canada does not desire to be a parasitic nation. She does not, and will not, wish to lean on anyone."

Some men who have not studied the subject have declared that Canada need not bother about means of defence, as the United States with her "Munroe Doctrine" will protect us.

As Homer Lee points out, "In the time of Munroe, it was impossible to foresee the changes mechanical inventions were to make in the political development of the world after his time.

No longer, as in Munroe's time, does a vast Atlantic separate this continent from Europe. Man's ingenuity has reduced it to a small stream, across which the fleets of European powers can cross in less time than it took Munroe to post from Washington to Boston."

He speaks in the same strain of the Pacific Ocean, and points out the imminent danger of a Japanese conquest of the Western United States, owing to the absolute commercialism of our neighbors to the south, and their neglect to take the necessary precautions of defence.

Assuming that some military and naval force be necessary to the existence and welfare of this country, we live in a too democratic age to permit a small number of our men who are sufficiently patriotic to spend their time, means, and possibly their life's blood to defend the others who sit at home in ease and—for a small price paid in taxes—enjoy the privilege of reading these things in the daily papers.

I do not wonder at the agitation for woman suffrage. Have they not as much right to vote as their brothers who refuse to take a man's part in being trained for the defence of their firesides?

Apart from all military considerations, every engineer who may have to do with organization of forces of men, or of a staff to direct any operations in the engineering profession, will find that the discipline which can be learned only in military life gives a very important advantage over his competitor who has neglected his opportunities along this line.

I know many will say "There is no time. All the time we can spare from study is taken up in music, or football, or cricket, or the gymnasium." These are all good, but to my mind, they do not compare with the military training which is as important as mathematics.

I am pleased to know that the University has one military organization, a field company of Engineers, which trains about one hundred men.

We should have every man in the University who is physically fit enrolled and given the opportunity of learning to obey before he is called upon to command and of obtaining the first principles of a training that will fit him to help to defend when necessary this Canada of ours, and to keep it for all time a part of the greatest empire the world has seen.

You may find, as do most engineers who stick to the purely professional side of the work, that the prizes in the profession are few and not very large when compared with responsibilities and the labor; but when all is balanced up, you will be able to say you have lived a man's life among men, and have done something that has left its mark on the development of your country, and have been of service to your fellowmen.

NOTES ON THE ELECTRO-METALLURGY OF IRON AND STEEL.

T. R. LOUDON, B.A. Sc.. A. M. Can. Soc. C. E.

Although the electric furnace has been used in the metal-lurgical industry for a considerable period of time. it is only within recent years that it has been possible to successfully produce pig iron and steel on a commercial scale. At the present day, the problem of mere mechanical operation seems to have been solved, so that it is only a question of efficiency and cost of electric energy as compared to ordinary fuels that determines the commercial value of the electro-thermic process for any given locality.

It will be the endeavor in this article to explain concisely the working of a few of the different furnaces used in the electro-thermic reduction of iron ore and the subsequent production of steel. It is, of course, understood that the accompanying drawings are extremely diagramatic. They will, however, serve very well the purpose of illustrating the theory embodied in each process. Should the reader care to go more fully into the question, there will be found throughout this article reference numbers. the key to which is given at the end.

Before discussing the various processes. it is extremely important to understand the part that electrical energy plays in the production of either pig iron or steel. In the ordinary process of smelting. the ore. together with limestone and coke, is dumped into the blast furnace, and as the result of certain chemical reactions that take place, pig iron is produced. Stated in as simple a manner as possible, the hot coke at the lower levels in the furnace meets a blast of air, and the ultimate result is that there is a formation of carbon monoxide gas. This gas, together with the carbon of the coke, reduces the iron from the ore. Now it is a matter of common knowledge that it would be impossible to bring about this reduction of the iron if the process were not carried on at a high temperature, and in order to attain this temperature, a certain percentage of the coke in a blast furnace is there for the purpose of producing heat. So then it is seen that the coke is present for two definite purposes: (1) To provide an agent with which to reduce the iron from its ore (2) as a source of heat. It is for this last purpose that electric energy is used: i.e., as a source of heat.

Working on the basis that the requisite heat is to be supplied by electrical means. if the cost of electrical energy required to give the same amount of heat as a ton of coke be figured out, it will be found that the price of coal for coking would have to be extremely high, and electric power very low before one could ever hope to replace the former with the latter. This calculation. however, takes no account of the fact that electric furnaces are far more efficient in their utilization of heat than furnaces using ordinary fuels. When these efficiencies are taken into consideration, it is found, roughly speaking, that with electrical power costing ten dollars per electrical horse-power per year, and coal at about eight dollars per ton, the electric reduction furnace may compete successfully with the ordinary blast furnace. These figures are, of course, the extremes in cost, but there are many localities where such conditions exist. (Reference No. 1 and 2).

As in the making of pig iron, so with the processes for making steel; the electric current is the means of providing heat. The ordinary furnaces for making steel, however, are far less efficient in their utilization of heat than the blast furnace, so that the electric steel furnace with its high efficiency is able to compete with some of the other processes under the ordinary commercial conditions. This is especially true where the electro-thermic process is used in conjunction with the ordinary processes.

Electro-Thermic Reduction of Iron Ore.

Roughly speaking, the reduction furnaces in use at the present time may be divided into two classes; those with vertical electrodes embedded in the charge; and those with the electrodes projecting into a crucible at the lower part of the furnace. Both of these types are known as "resistance furnaces," due to the fact that the heat is generated by the resistance offered to the passage of the current through the charge.

The simplest form of furnace is that shown in Fig. 1. The carbon electrode A is suspended by some regulating device in

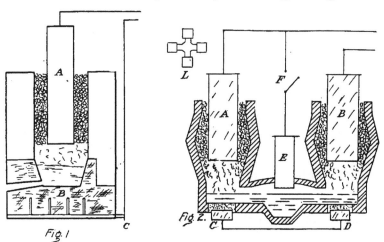

Fig. 1   Fig. 2.

the charge of ore, coke and limestone. The crucible walls and the hearth are made of a carbon paste, in which is embedded a second electrical connection, C. If, now, the circuit be closed, the current in passing from A to B, or vice versa, will, owing to the high resistance of the intervening charge, develop the neces-sary heat requisite to attain the desired temperature.

A furnace of this type was successfully used in the experi-ments carried on at Sault Ste. Marie by the Canadian govern-ment under the direction of Dr. Haanel. While these experi-ments were entirely satisfactory, it was pointed out on their completion that there would have to be some modifications made in order to allow the charge to feed regularly and to utilize the gases as they escape from the top of the furnace. An attempt was made at the time to make use of these gases by introducing an air jet into the charge near the top, thus burning the gases, but it was found that the charge became sticky, and was inclined to "hang"—a very bad fault. (Reference No. 3).

Fig. 2 represents a type known as the Keller furnace, in-vented by Mr. C. A. Keller of France. A and B are two carbon electrodes. Embedded in some conducting material in the hearth are two electrodes, C and D, connected together as shown. To begin operations, A and B are lowered and the charge packed around them. At first, the path of the current will be from A down the charge, across from C to D and up the charge to B, or vice versa. As soon as the iron covers the hearth, however, the current instead of passing around the shunt from C to D, as described, will take a path through the molten metal, thus tending to keep it warm. Sometimes, though, the bath may cool at the centre. If this occurs, the auxiliary electrode E is brought into play by closing F, part of the current then passing through the central portion of the bath, giving a greater intensity in that locality.

This furnace may be built with a plurality of hearths, as in-dicated at L, Fig. 2, each of the small squares representing a stack with its vertical electrode. By this arrangement the elec-trodes may be arranged in pairs in parallel if so wished, thereby giving a better system for regulation. (References Nos. 4 and 5.)

Leaving for the time being the discussion of these vertical electrode types, there are shown in Figs. 3 and 4 two furnaces of the second class indicated previously.

Fig. 3 shows diagramatically the construction of a furnace which is the outcome of a number of experiments carried on at Domnorfvet, Sweden, by Messrs. Grönwall, Lindblad and Stal-hane. A number of different furnace shapes were tried until finally the construction indicated in Fig. 3 was arrived at.

The upper portion of the furnace is in appearance very much the same as an ordinary blast furnace. This shaft is built on a large crucible or melting chamber, B, projecting through the roof of which are three electrodes, one being shown at A. At the top of the shaft, there is, of course, a cover with a charging bell very much the same as on the present-day blast furnace. The gases coming off the charge at the top are led out at C to a dust catcher. D. Cleansed to a certain degree, this gas is blown by means of a blower. E, back into the crucible at F.

It will be noticed that the tuyeres by means of which the gas is blown into the crucible slant upward. This is for the purpose of allowing the gas to impinge against the crucible roof, thereby cooling it. This does not in any way lower the efficiency of the furnace, the heat absorbed by the gas in cooling the roof. being given back to the charge as the gas flows upward again. Indeed it would appear to be a great factor in raising the efficiency, since the heat absorbed by this gas in the crucible would have otherwise been most likely lost by radiation. The tuyeres

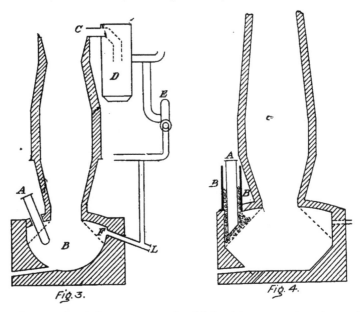

Fig. 3.   Fig. 4.

(three in number) by means of which this operation is performed, have peep-holes, as shown at L, so that the operator may judge the roof temperature, and accordingly regulate his jets of gas.

In this process there are no electrodes embedded in the walls of the furnace, the current passing between the three projecting carbons, thus giving intense heat in the crucible.

The Frick furnace, shown at Fig. 4, is somewhat similar in construction to the last furnace. A vertical shaft is superimposed upon a crucible, and through the roof of this lower melting chamber there project two vertical carbon electrodes, one of which is shown at A. These electrodes are suspended in casings or shafts, shown by heavy lines at BB.

The distinguishing feature of this furnace is the fact that the electrodes are packed around loosely with a reducing agent such as coke or charcoal. This coke or charcoal forms a protecting layer around the portion of the electrode in the crucible, and also, intervenes between the electrode and the charge as indicated. (The dotted lines represent the slope of the charge as it works down from the shaft).

The purpose of providing this covering of coke or charcoal, as the case may be, is to protect the electrodes from being oxidized away. It can readily be seen that the carbon of the electrode would itself be oxidized if there should happen to be quantities of ore around the electrode. The reducing agent, however, protects the electrodes from being oxidized by entering into the reactions itself, and also since there is an intervening layer between the electrode and charge, there will be far less wear on the electrode, as the charge slides down.

There is also provision made in the furnace to carry the gas from the top and blow it back into the crucible if it is so desired. (References 6 and 7).

Electric Steel Furnaces.

Furnaces in which the refining action necessary for the production of steel may be carried on can be divided into two classes: (1) Induction Furnaces; (2) Arc Furnaces. The latter class can be sub-divided into furnaces in which the arc is formed between the electrodes above the bath; and those in which the arc is between the electrodes and the bath.

Furnaces of the induction type are constructed in such a manner as to make use of a well-known electrical principle which may be stated to suit this particular purpose as follows: If an alternating current be passed through a coil of wire known as the primary coil, and if there be surrounding this coil a second

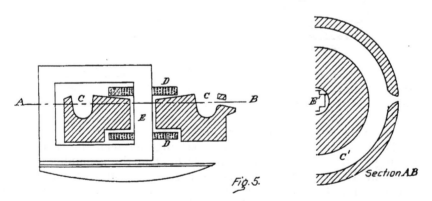

Fig. 5.   Section AB

independent coil, there will be induced in this secondary coil, as it is called, an independent alternating current. In order to intensify this effect it is usual to wind the primary coil around an iron core.

Fig. 5 represents two sections of one of these induction furnaces. At D and D are two coils of wire wound on an iron core. E. Surrounding this core is a circular channel, sections of which are shown at CC. If a horizontal section, AB, be taken, the channel will appear as at C', surrounding the core E'. (Merely half of this section is shown).

If now the channel CC be filled with a charge of molten iron, there will be formed around the core and coils DD a closed coil of one turn; so that if an alternating current be passed through the coils DD, there will be induced in the molten metal another independent current. It is this induced secondary current that furnishes the heat in the charge forming the ring CC.

There are a number of these induction furnaces being used for the manufacture of high quality steel. The particular type shown diagramatically in Fig. 5 is known as the Frick furnace: but there are several other types, outstanding among which are the Kjellin and Röehling-Rodenhauser. In the Kjellin furnace, there is merely one long vertical coil around the core instead of the two flat coils as in Fig. 5. There are, of course, other differences, but the explanation given for Fig. 5 will apply to the Kjellin furnace, which has a channel around the primary coil for the charge. (References 5 and 11).

The Röchling-Rodenhauser furnace, while it is of the induction type, has a feature which is extremely novel. This furnace may be said to have two secondary currents, one of which is the usual current induced in the ring of metal as explained before. The other current has its source as follows: Around the primary coil is placed a secondary wire winding, the terminals of which are led to electrodes embedded in the walls of the channel for the metal. Thus there is induced in this secondary winding a separate current, which in passing through the bath of metal adds to the heating effect of the current already induced in the ring of metal itself.

Unfortunately, it is impossible in such a synopsis to go more completely into the details of these furnaces, but there will be found in References 5, 7 and 8 abundant information regarding the three types mentioned.

Passing to the arc furnaces, at Fig. 6 there is illustrated the principle of the Stassano furnace. Projecting through the walls of the furnace are three electrodes, as shown in the cross section. Between these carbons is formed an arc, A, the heat from which is radiated to the bath of molten metal B. The furnace, which is inclined, is given a rotary motion about its upright axis, thus causing a stirring of the charge. The regulation of temperature in such a furnace is under very easy control. (References 5 and 9).

Considering next furnaces in which the arc is formed between the electrodes and the bath, there are two types which seem to stand out pre-eminently, viz., the Héroult and the Girod furnaces, the principles embodied in both being easily explained.

Fig. 7 illustrates the method of making steel in the Heroult furnace. Suspended through the roof of the furnace are the electrodes A and B. The charge lies on the hearth and these electrodes are lowered till an arc forms between them and the bath. The current passes down one electrode, arcs to the bath, goes through the bath, arcs to the other electrode, and out again.

The upper cross section, Fig. 7, merely shows diagramatically that the furnace is so constructed to allow it to be tilted forward when the charge is to be withdrawn. The electrodes are held in position by braces attached to the furnace as indicated. (Reference 5).

The Girod furnace, while it is in outward appearance very much the same as the Héroult, differs from it in this respect.

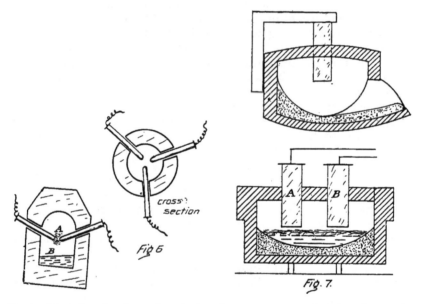

cross section

Fig. 6

Fig. 7.

Embedded in the hearth of the furnace are suitable electrodes, and the current, instead of passing in and out from the top carbons as in the Héroult furnace, has a path from the upper electrode, or electrodes, through the bath to the connections in the bottom of the furnace. Thus instead of the upper carbons being in series, as in the Héroult system, they are in parallel. In actual operation, the current in the Héroult furnace passes along the slag, thus providing a very hot blanket as it were. In the Girod system, the current in passing through the bath itself gives, it is claimed, a method more suited to working from a cold charge. As against this is the counter-claim that furnaces with electrodes embedded in the hearth are troublesome to repair. Be all this as it may, there is the incontrovertible fact that both systems are now in extensive use. (See reference 10).

In conclusion, it is interesting to note that not only are these steel furnaces being used to displace the old inefficient crucible steel plants, but there are notable examples where the larger fields have been entered. Quite recently the Prussian state railways have been buying rails from a firm manufacturing steel by the Röehling-Rodenhauser system. These rails when tested chemically and physically were of a much higher standard than the specifications called for. In the United States the Illinois Steel Company has lately been refining molten steel from the acid Bessemer process by means of a Héroult furnace. There are also other instances where the electric furnace is being used in combination with either the Bessemer or Open Hearth processes as will be seen in Reference 10.

In Reference 10, there will be found a tabulated list of steel furnaces in use throughout the world. It will be noticed that in most cases alternating current is used—mostly single phase— although direct current is sometimes used. The reduction furnaces described use an alternating current.

As to the actual working of these electric furnace processes, it may be pointed out that the regulation of the metalloids in the desired metal follows the theory of the ordinary processes, the main difference being that it is possible to maintain very much more basic slag.

SUMMARY.

The advantages of the electric furnaces, reduction and refining, may be stated as follows:

1—Efficient production of heat.

2—Ease of regulation as regards temperature.

3—On account of the high temperatures attainable, it is possible to maintain a very basic slag.

4—Absence of injurious gases in steel working processes.

5—The possibility of using charcoal as a reducing agent for making pig iron.

KEY TO REFERENCE NUMBERS.

No. 1—The Electric Furnace, Its Evolution, Theory and Practice (Alfred Stansfield).

This book contains a very concise and clear outline of the historical development of the electric furnace. There will be found in it a discussion of the relative costs of electric energy and fuel for producing heat.

On page 250, vol. 13, of the *Canadian Engineer* will also be found a discussion of these relative costs by the same author.

2—"Metallurgical Calculation," vol. 1 (J. W. Richards).

The methods outlined in this book are extremely good.

3—Report on experiments at Sault Ste. Marie, 1907 (Eugene Haanel).

These experiments were carried on by the Canadian government.

4—Application of the Electric Furnace in Metallurgy (A. Keller).

Journal of Iron and Steel Institute, 1903, vol. 1, page 161.

5—Report of the Commission appointed to investigate electro-thermic processes of smelting iron ore and making of steel (Eugene Haanel).

This is the most comprehensive report ever made on electric processes up to 1904, and is even yet invaluable.

6—Investigation of an electric shaft furnace, Sweden (Haanel).

7—Recent advances in the construction of electric furnaces, etc. (Haanel).

The last two reports published by the Canadian government are very complete.

8—Electro-chemical and metallurgical industry, vol. 6, pages 10 and 458.

Good descriptions of the Röchling-Rodenhauser furnace and its operation.

9—Same periodical as 8, vol. 6, page 315. This is an exhaustive article by the inventor of the Stassano furnace.

10—Stahl and Eisen, March 23, 1910, page 491.

This gives a tabulated list of furnaces in use. See also No. 7 for same thing.

11—A description of a Kjellin furnace will be found on page 397 of vol. 3, journal of Iron and Steel Institute.

PROSPECTING IN THE COBALT DISTRICT.

H. L. BATTEN, '11.

Prospecting covers a great many operations, from the search for, and staking of claims, to actual mining work underground. This paper does not deal with the search for claims, but is an attempt to give some idea of the work to be performed after the claim is staked, for the purpose of determining the value of the property, position and value of veins, etc. We are still left with surface prospecting and underground work to be considered, as both have to be performed in practically all cases before a property may rightly be called a mine. There can be no doubt that a thorough prospecting of the surface before commencing mining operations is of the utmost importance, and should be taken seriously. In spite of this fact there are quite a number of properties around Cobalt and South Lorraine with a shaft down 100 feet or more on a one-inch vein of calcite carrying no values, while not a shovelful of dirt has been moved on the other parts of the property. The cry is, "We have no money to spend on the surface, and besides, all the veins do not show on the surface." This is no doubt true, to a certain extent, but if we try to make out a list of paying mines around Cobalt that had no showings on the surface, we shall find there are very few.

Again, a great number of the properties that are supposed to have been prospected have never been prospected thoroughly. The work has been done in a haphazard manner by the original owners, or by men hired by them, as assessment work. The idea has been to put in so many day's work, and no successful prospecting can be carried out in this manner. For the work to be thorough, some system must be adopted, not necessarily to be blindly followed, without modifications, but to serve as a guide. Unless some method is adopted, trenching will not have progressed very far before the question "where to put the next trench" will arise. Again, if the positions of the trenches be determined before work commences, it will probably soon be found that a trench could be run, with much less work, a few feet to one side, and be of equal value for prospecting purposes.

The idea of surface work is to lay bare the bed rock, so that it can be examined. This is usually done by digging parallel and perpendicular trenches, thus dividing the property into a number of squares or rectangles. The distance between the trenches must be determined by the circumstances. The closer the trenches the less likelihood of veins being missed, but a great number of things have to be considered. If the surface covering is deep it may not be advisable to run trenches less than 200 feet apart on account of the expense. On the other hand, the Nipissing Mines intend to remove all the surface covering by hydraulicing.

It is obviously impossible to lay down any rules, as the distance between trenches must be determined only after considering all the circumstances; but on most properties probably about 60 feet apart would be a suitable distance. The distance apart will probably not be the same all over the property. It may be found after putting down a few trenches that, while part of the property should be prospected very carefully, the remainder may not show indications of being worth going to much expense over. No hard and fast rules can be laid down.

As regards the direction of trenches, the usual way is to run north and south, and east and west trenches. When no veins are known on the property, and the surrounding properties have not been prospected, these directions are as good as any, and are most convenient. However, it is a somewhat generally accepted rule that veins in proximity to each other are likely to be approximately parallel. The best direction, therefore, to run the first trench is perpendicular to a vein that has already been uncovered. There is almost sure to be such a vein, marked by a discovery post, and a start should be made at that vein. On a property prospected by the writer this last summer, one vein was known before commencing to prospect, striking approximately north and south. By running trenches perpendicular to this vein four others were uncovered, their strikes varying from north and south to north 50 degrees east. Thus the strike of a vein probably is an indication as to how nearly veins may be expected to run; but veins are extremely erratic, and are almost as likely to be found running in any other direction.

The first thing to be done in prospecting a property is to find out everything possible about it; assuming, of course, that everything possible has already been found out about the company working it. The boundary lines should be carefully and clearly marked out or a great deal of time will be wasted hunting up corner posts. The whole property should be gone over carefully, in company with a man who knows the ground, if possible. If time permits, a rough sketch map should be made, showing the relative positions of any veins that may be known, outcropping rock, swamps, or other places difficult to prospect, and as much of the geology as can be determined. If the timber has been cleared off, this work should not take long, but if the bush is thick a week probably can be profitably spent in this manner.

It is necessary to be fairly well acquainted with the property before deciding on the best way even to start prospecting, cr where to start, so that the work may be put in to the best advantage. A little care and forethought can save money in prospecting as in everything else, and no start should be made until the whole claim has been carefully examined. There is always some doubt where to make a start. If the vein on which the discovery was made has not been stripped, three or four men (not more than four men can work to advantage, stripping a vein that has been uncovered in only one place), should be put on this job while the claim is being examined. This vein should be stripped for a sufficient distance to give a good idea of its value. If the indications are promising, it (and all the veins) should be stripped as far as it can be followed. However, as soon as the approximate strike has been determined, more men can be taken on and a trench started perpendicular to the vein; or approximately so.

The property will probably be divided naturally into several sections. The goal to be aimed at is to show results in the shortest possible time, and work should therefore first be done where the indications are best; or if there are no indications worth considering, a start should be made where the most bed rock can be stripped in a day. The section containing the vein should be the first to be prospected, and an attempt must be made to pick up any parallel veins there may be.

Contacts must be considered. If diabase outcrops in one place and Keewatin rocks in another, they should be joined by trenches, and the contact located. The reason for this is that there is a likelihood of veins occurring at or near the contact. In the case of the five parallel veins referred to previously, the first was at the contact of diabase and conglomerate, while the other four were all within 250 feet of the contact.

In working over the ground in sections however, care must be taken not to lose sight of the general method which has been adopted. Before the work is complete the trenches in the different sections have to be connected up, so that each trench runs from one boundary line to the line opposite. This may not be possible on account of a swamp or drift too deep to make trenching practicable, but should be carried out whenever possible.

Where heavy trenching is encountered it should be left until all the light work has been done, unless there is some special reason why it is desirable to put down a trench in that particular place. When to leave a trench (temporarily, of course), is a matter of judgment. As a general rule it may be said that all trenches requiring timbering should be left till the lighter work is finished. In the spring no attempt should be made to put down wet trenches—it is almost hopeless to try and make sure of missing no veins and yet make good progress with wet trenches. Also, it is impossible to make good footage with heavy or wet work, and it is most important to make a good showing at the start.

Before settling down to prospect thoroughly, section by section, the whole claim should be gone over by running a few preliminary trenches. This is especially advisable if outcropping rock be scarce. As accurate a map as possible should then be made, and on this map everything that is known should be marked. This map should be carefully made and brought up to date at least at every month-end. The positions of all trenches should be accurately shown, and the positions of all veins and test-pit marked. When a few trenches have been put in, the topography should be sketched in, and the elevations of all important points shown. The only instruments required are a compass and chain, and an aneroid barometer, for obtaining the relative elevations. This map is useful for several purposes. It can be seen at a glance what ground has been prospected and what remains to be worked, and the required positions of trenches to connect with trenches already dug can be determined. If the bush is thick these cannot be lined in by sight. Also reports on the progress made are greatly simplified by the possession of such a map. A tracing can be made each month-end, with each month's work shown in different colors, and handed in with a short description of any discoveries made. Such a report conveys more information and creates a better impression than a lengthy written account of the work performed.

There are a few points to be noticed regarding the details of the work, and one of these is the best widths for trenches. It must be remembered that a slight increase in the width of trenches means a considerable increase in the amount of dirt to be shovelled. Economy therefore, demands a narrow trench. In a narrow trench, however, it is very difficult to clean off the bed rock thoroughly. Three feet in width should probably be considered a minimum, and this width is sufficient only for trenches up to about three feet deep. A trench five feet deep should be four feet or four feet six inches wide on the surface, and deeper trenches, even if no timbers are required, should be wider still. The best width depends to a certain extent on the banks, so it is impossible to lay down any rules, but the general tendency is to dig trenches too narrow, which means difficulty in examination, and a consequent likelihood of veins being passed over unnoticed.

As bed rock is stripped it must be very carefully cleaned off. If the surface is dry sand or gravel, this is best done by brushing. If the soil sticks to the rock, as clay, for instance, does, the rock must be washed, and washed thoroughly. Care must be taken not to fill up the cracks and crevices with mud, as it is these which have to be most carefully examined. It is better not to wash off the rock unless absolutely necessary, owing to this trouble caused by the inequalities being filled with mud.

Trenches cannot be prospected too carefully, as especially in diabase, it is often extremely difficult to recognize a vein. The bed rock should be most carefully watched for any change in texture, as the rock near a vein is often finer grained than the surrounding rock. Too great care cannot be exercised, and every place where there could possibly be a vein should be shot out. The number of veins which have been found in old trenches, and which had previously been missed is startling.

With even the greatest care some small veins will probably be passed over, and a vein less than one inch in width in this district may be extremely valuable.

Having found a vein, the amount of work that should be done on it depends on the vein itself. It must at least be stripped for some little distance, and a test pit sunk deep enough to get below the weathered surface. If the indications are in any way promising it should be stripped and opened up as far as it can be followed, and the trench should be sufficiently wide to allow the vein to be shot out without having the banks fall in on it. A considerable amount of shooting is unavoidable when opening up a vein, but as much drilling as possible should be avoided, since it is extremely difficult to get a hole down to any depth owing to the number of joints and seams usually found near a vein, and even under the best conditions, drilling is a slow and expensive job. A vein can usually be opened up by exploding powder in the fissure (providing the loading and tamping be properly done). More powder will be used than with drilled holes, but the saving in labor more than repays for this.

Before reporting any find, care should be taken to be absolutely certain. If in very small pieces, bismuth and silver are easily confounded. When looking at museum specimens there appears to be little similarity between the two, but when silver is expected and a speck of bismuth encountered, it is very easy to be misled. in fact, it is often absolutely impossible to be certain without a chemical test. Cobalt bloom and red iron oxide are frequently confounded. Unless there can be absolutely no doubt a chemical test should always be made before any find is reported.

When the surface work is completed there are two methods of prospecting underground to be considered—the ordinary mining methods, of shaft-sinking. drifting and cross-cutting; and the diamond drill.

It can safely be said that the diamond drill has not been a success in this district, at least as far as the search for, and testing of veins is concerned. There are two principal reasons for this failure; first, as the drill passes through the vein the latter breaks up into powder, and no core of the vein-matter is obtained. Secondly, even when a sample of the vein is obtained, the information gained is of little value. Only one point in the vein is sampled, and while there may be no values or very low values just at that point, other parts of the vein may carry high values. Sufficient data to form any idea of the value of a vein cannot be obtained with the diamond drill.

Ordinary mining methods apply to shaft-sinking, etc.. whether the property is a prospect or a mine. It is obviously impossible in this paper to go into such questions as to when a plant should be installed. or when to work by hand. Around Cobalt this question has been solved to a great extent by the advent of the power companies. In South Lorraine it still exists, although they are expecting to have electric power there this fall (1910).

It must be remembered that in prospecting, as in all mining ventures, no two propositions are alike. The work to be done depends upon so many things; the property itself, amount of money to be spent, the length of time in which to perform the work, and a host of other circumstances have to be taken into consideration. There is room for differences of opinion at every turn, and there is no pretence made that this paper is in any way complete, or even representative, it is simply a collection of a few opinions formed during a couple of summers' prospecting in Cobalt and South Lorraine.

GAS AND OIL ENGINE RESULTS.

Thermodynamic Laboratory, University of Toronto.

W. W. Gray, B.A. Sc.

A series of tests was conducted by a party of fourth-year students, Messrs. Schwenger, Stroud and Thompson, on the 9 B. H. P. Fielding and Platt gas engine, to determine its gas consumption per B. H. P. per hour at various loads.

The engine is of the single cylinder, single acting, horizontal, hit-and-miss governor type, and is of standard English design throughout. The results obtained are very satisfactory for an engine of this small power.

Results of Tests.

City illuminating gas.
Diameter of cylinder, 7 inches; stroke, 14 inches.
Compression, 90 pounds per square inch.
Ignition, electric, by trip magneto.

No. of test	1	2	3	4	5	6	7	8	9	10
Date	Jan 25	26	26	26	26	26	27	28	28	28
Duration (minutes)	40	30	30	30	25	30	25	30	15	30
R.P.M.	220	220	219	219	219	220	218	218	218	218
B.H.P.	0	1.19	2.03	3.04	4.05	5.08	6.04	7.06	8.06	9.07
Gas per hour, cu. ft.	51.2	91.5	86	93	114	121	150	153	166	182
Gas per B.H.P. hour cu. ft.		76.9	42.4	30.6	28.1	23.8	24.8	21.7	20.6	20.1
Ratio — Gas	1	1	1	1	1	1	1	1	1	1
Air	9	6	10	10	10	12	10	10	10	10

This engine is provided with a special vaporizer and valve gear, to allow the use of kerosene, the vaporizer being kept at the proper temperature by means of a kerosene torch.

Ignition is by the hot tube method, the tube being heated by the vaporizer torch.

A series of tests at various loads was conducted on this engine by Messrs. Black, Manning and Taylor, a party of fourth-year students.

Results of Tests.

Fuel, kerosene.

Compression, 90 pounds per square inch.

Ignition, hot tube.

No. of Test	1	2	3	4
Date	Feb. 17	17	17	17
Duration in minutes.	30	30	30	25
R. P. M.	213	220	221	212
B. H. P.	. . .	1.51	3.03	7.75
Kerosene per hour (lbs.).	1.937	2.562	3.687	6.187
(Including Torch)				
Kerosene per B.H.P. hr. (lbs.)	. . .	1.7	1.22	.8

Diagram 1, taken from the engine when it was operating on city gas, is a characteristic card for a gas engine.

Diagram 2, an oil card; is a very good diagram for an oil engine; very few similar cards were obtained during the tests.

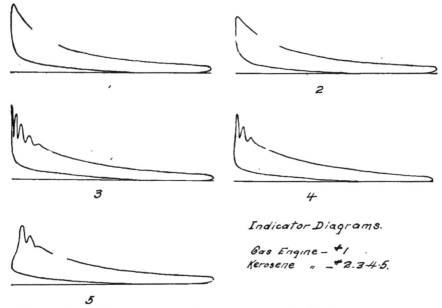

1

2

3

4

5

Indicator Diagrams.

Gas Engine — #1
Kerosene „ — #2-3-4-5.

To produce this diagram, the treatment of the kerosene must have been under ideal conditions, with respect to temperature of vaporizer, proportion of air and vapor in the mixture, and temperature of ignition tube.

Diagrams 3, 4, and 5 indicate too rich a mixture, a condition difficult to overcome in oil engine work.

NOTE ON THE UTILIZATION OF ATMOSPHERIC NITROGEN.

SAUL DUSHMAN, B.A.

During the last few years a large number of processes have been devised for the utilization of atmospheric nitrogen. The motives for these processes have been two-fold: both in fertilizers and in explosives nitrogen forms a very important constituent. Nitre or Chili saltpeter has been the substance chiefly used for these purposes in the past: but the nitre beds are gradually becoming exhausted, and the world has come to realize that some other source of fertilizers must be sought for if its food supply is to remain assured. At the same time there exists a continually increasing demand from the manufacturers of explosives, for a concentrated nitric acid.

As far back as 1776, Cavendish, the famous English physicist, discovered that an electric spark passed through a moist mixture of oxygen and nitrogen produces nitric acid, and at least three processes are in commercial operation which are based upon this reaction. In the process devised by Birkeland and Eyde in 1904, a 5,000-volt alternating current is passed between water-cooled copper electrodes and by means of a magnetic field at right angles to the latter, the arc is spread out into a disc of over two meters in diameter, thus causing almost the whole volume of air in the furnace to be raised instantaneously to a very high temperature. About one per cent. of the air passing through the arc is thus converted into oxides of nitrogen, which are then washed out by passing the gases from the furnaces through a series of towers. There is obtained in this manner a 50 per cent. nitric acid, which is converted into calcium nitrate by neutralizing with lime. The latter is used as a fertilizer directly. The furnaces used by Birkeland and Eyde were originally of 500 k.w. capacity, but subsequently they were replaced by 800 k.w. units, and at the present time it is the intention to replace these by 1,600 k.w. furnaces. In the first commercial installation at Notodden, Norway, only 1,500 k.w. was used; afterwards this was increased to 40,000 k.w. During 1908, the first year of operation, the total income was $536,000, and the net gain, $134,000. Plants have been established at numerous other places in Norway, where water-power is available, and according to most recent reports the Norwegian industry of manufacturing air nitrates is undergoing rapid extensions involving the expenditure of nearly $15,000,000. Not only calcium nitrate, but also more concentrated nitric acid, nitrate of ammonia, nitrate of potash, as well as sodium nitrate are being manufactured at these plants. One of the principal reasons for the success of the process in Norway is undoubtedly the small cost of power which is said to be available at about $5 per horse-power year.

Another process based upon the same fundamental principle is that developed very recently by the Badische Anilin-und Soda-Fabrik. "In this process a continuous arc is produced in a long tube by first bringing electrodes together and then gradually moving them along the tube while the other remains fixed at one end of the tube. The current of air, instead of being passed through this arc, is passed around it through the tube by being forced in at an angle to the main axis of the tube. It is said that the arcs used in this process vary from 35 to 50 feet or more in length, and are maintained continuously for days or even months at a time."

Besides these two processes which are both in successful commercial operation, there is a third process for the oxidation of atmospheric nitrogen which has been devised by H. and G. Pauling, and is working on a large scale near Innsbruck in Tyrol. The electrodes are curved like the electrodes of the so-called horn lightning arresters. The arc is started at the narrowest part between the electrodes by means of a special lighting device. The air current passed through the arc blows it out to a considerable distance and thus increases the volume of air heated to the extremely high temperature at which combination occurs. "The twenty-four furnaces at present installed in this plant have a total capacity of 15,000 h.p. Two other plants, each of 10,000 h.p., for carrying out the same process, are in course of erection, one in Southern France, and the other in Northern Italy."

The process of Frank and Caro for the utilization of atmospheric nitrogen is totally different from any of the above methods. Nitrogen is passed over heated calcium carbide and the result is the formation of a substance having the formula $CaCN_2$ and known as cyanamide. Investigations at numerous agricultural stations have shown that it can be successfully used as a fertilizer, and accordingly a large number of plants are being erected both in Europe and America for its production. The United States Cyanamide Company has a 5,000-tons works at Niagara Falls, Ontario, and is also building another plant in Tennesee. Cyanamide is also of interest on account of the number of interesting derivatives which it is capable of yielding when treated with different reagents. An important reaction is that with steam, leading to the formation of ammonia, which may be subsequently converted into ammonium sulphate.

Still another method for the utilization of atmospheric nitrogen has been devised recently by F. Haber and patented by the Badische Anilin-und Soda-Fabrik. A mixture of nitrogen and hydrogen in the required proportions is maintained at a constant high pressure and heated in presence of a catalytic reagent (such as finely divided iron or osmium) to a temperature which varies between 400° and 800° C. Only about 8 per cent. of the mixture is converted into ammonia, but the power necessary for the compression and circulation of the resulting gases is very small, and it appears very likely that in the near future this process will be exploited industrially.

TECHNICAL EDUCATION

A Synopsis of the Discussion at a Meeting of the Engineers' Club, October 13th, 1910.

Dr. Galbraith, *Dean of the Faculty of Applied Science and Engineering.*

Mr. Chairman and Gentlemen:—The object of having a discussion on this subject by the Engineers' Club is for the purpose of presenting some of our views to the Royal Commission on Technical and Industrial Education. Thinking that the members of the club might have some ideas on this subject which they would like to present. I asked the Commission if they would care to receive our views. They assured me that they would welcome such opinions, and would be glad to receive them upon their return to Toronto for the examination of witnesses.

To-night I am asked to open the discussion, and do not propose speaking longer than is necessary to present the question and to elicit the opinions of the members of the club who are present.

The term "Technical Education" is very broad. so broad indeed, as to be almost vague. The Faculty of Applied Science in the University of Toronto, covers only one branch of the subject, and it may be the duty of the Commission to study the question in its most general sense.

The name is almost a misnomer, as usually applied. The word " technical " really means, "relating to the art of making things or doing things": for example, a trade education in the methods of making things or the doing of work by manual labor is technical in the original sense of the word. The expression is now applied generally, to training in practical science.

As an illustration of some of the difficulties in technical education, if you turn to the curriculum of the Toronto Technical School, you will see that very nearly the same subjects are taught in the evening classes as at the University in the Faculty of Applied Science. A stranger, not knowing the circumstances could not make any distinction. Now, the difference is this: the objects of the Faculty of Applied Science are well defined: they are limited by the nature of the professions to which the training is directed. and consequently, it is comparatively easy to construct the curriculum. But in the case of the municipal technical school. the students come from all sorts of positions

and trades. They often have no definite object in view. One de-
sires one thing, another wants something else. It is hard to determine what to include and what to leave out, and this is due to
the fact that the requirements of the students vary. They cannot be divided into graded classes as are the students in the
Faculty of Applied Science. They require to be treated as individuals rather than in classes.

Another difficulty is this: in the university (I am not speaking of Toronto only, because the principles apply generally), the
matriculation, or entrance examination, whatever its weaknesses
may be, accomplishes one object. The students know at the beginning, and the teachers know, just what the qualifications of
the students are. The instructor knows where to begin, and
this in itself is a great advantage. In the evening classes of the
Technical School the situation is quite otherwise. Satisfactory
grading by means of an entrance examination is difficult   or
impossible. These difficulties in the case of the evening classes
came under my observation when a member of the board   of
directors some years ago.

The teaching of trades in the municipal school in a place
like Toronto is rendered difficult by the great number of trades
to be considered; the cost would be enormous. Again the problem is made more   complicated by the   attitude of the   labor
unions. The representatives of the unions at the time I speak of
were in favor of teaching   practical science, , but opposed   to
teaching trades. The problems to be solved in the case of   the
Toronto Technical School will be the problems of the municipal
technical schools in the majority of Canadian communities.

There is a noteworthy movement that is taking place   in
America, viz., the teaching of the trades by the people who are
most interested in them. This seems only common sense. It
is being done by the   larger   corporations; the Grand   Trunk
Railway, and the Canadian Pacific Railway have begun this work
in Canada. They have their own schools for their apprentices,
who receive instruction in both the theoretical and the practical
sides of the trade, and the results are very encouraging. The
instruction books used by these corporations are very well adapted to their purpose. While a man is working on one machine,
he is given a blue print describing the machine to be taken next.
He studies this, and is able to answer theoretical questions relat·
ing to the machine by the time he is moved on to it. This plan
is being used also, by the Baldwin Locomotive Works, in
Philadelphia, and in many other large plants. It is adopted by a
number of the largest corporations, and is done for purely business reasons. There is no philanthropy entering into the question, for corporations have no souls, and only work for the divi-
dends. In these corporation schools there are two classes   of
students as a rule, viz., the men from the universities and the
men with only a common school education. The trade educa-

tion of these men, while extremely practical in both cases, is different, for the reason that they are needed for different purposes.

Much the same system obtains in Germany. The cases of which I have heard are in the manufacturing works in which applied chemistry is the fundamental science. The laboratory is located in the works. The best trained men are taken from the technical high schools and placed in these works, where they are trained in the methods of the company. It seems that where the employee is trained by the employer the result is mutually beneficial. It is not like the training of a group of men for something they may never follow afterwards.

There is not much respect shown by students for a teacher, whether in theory or practice, unless he is an expert in what he teaches. If we are to make any progress in the teaching of trades, we must have men who are skilled in the trades they propose to teach, and these men are what the Grand Trunk and Canadian Pacific Railways have. They have also practical men who can teach the necessary theory.

To obtain teachers with suitable qualifications will prove one of the greatest difficulties in teaching trades. I believe some teachers can be obtained in Canada. I know that among the young men who enter the Faculty of Applied Science there are some expert mechanics, and these men when they graduate may be used as teachers, since they would be equipped to teach both theory and practice, and could hold the respect and interest of the pupils. But not every graduate in mechanical engineering could do so, because as a rule, when he graduates, he is neither an engineer nor a mechanic. The difficulties of obtaining teachers for trade schools will naturally be greater at the initiation of the work than afterwards; demand will doubtless encourage a supply.

I have now touched upon a few of the leading points as they occurred to me. I have not gone into details, but have attempted rather to present the subject to the Engineers' Club in such a manner as to suggest questions for discussion.

Mr. R. W. King, of *Robert W. King and Co., Architects and Engineers, Toronto.*

I would like to emphasize one point made by Dr. Galbraith, which is, the absence of localization in the different industries. This has been one great difficulty confronting technical education.

I had an opportunity of seeing this in connection with the trade of knitting. There are certain centres given to this trade. In England it is Leicester. Weaving had a centre at Nottingham, while in Germany there were other centres for these trades. In Leicester and Nottingham and in some of the other centres, trade schools and technical schools are to be found, because there is where the industry flourishes. The government makes grants to these schools with a view to an increase of the staff. Germany set the example by instituting an immense school where trades desired by the community were taught. Its students were taught how to do their work intelligenly. This was the sort of education that England found it necessary to adopt, and they sent their teachers over annually to get pointers. I have no experience outside the fabric industry, but I would be especially interested in seeing the experiment tried in that line.

I suppose the same would apply in other parts of the country where these centres exist. In Ste. Hyacinthe, Quebec, there is a centre for woollens and knitted goods. In Ontario we have Paris. Now, if we are going to follow the example of the Old Country, it is in such places that the technical or trade schools should be placed, and maintained by the government, I believe it would be to the advantage of the country to centralize the trade. You could not hope for much success in Toronto, I fear, because of the great variety of trades being followed.

C. H. C. Wright, *Professor of Architecture, Faculty of Applied Science.*

Dr. Galbraith has given you the academic side of this question, and I do not think that I can add anything to what he has so ably said. I believe that if the Manufacturers' Association deem it necessary to place their notes before the Commission on this subject, then the members of this club should also present their views; and I would like to hear what difficulties exist to you. What are your difficulties as practising men? Do your mechanics, machinists and artisans understand your instructions? Do they read the drawings, or do they blunder through them, making mistakes that cause endless trouble for you and for themselves? I would like to hear what these troubles in handling men arise from. I am not prepared to say that we have missed the cue, educationally, although there may be improvements necessary. We should know what the real difficulties are, the existing conditions, so that we can work together to clear them out of the way. I believe you men who are out in the practising world are the ones who have met the difficulties, and these are the things that I believe should be brought before the attention of the Commission.

T. Aird Murray, *Consulting Municipal and Sanitary Engineer, Toronto.*

I have been asked to give in a few words my ideas as to how far and in what way methods of education can be outlined to produce what is technically called a "Sanitary Engineer."

I understand that views expressed to-night are to be laid before a commission whose business is to report upon technical education.

My difficulty is at once apparent. I am not sure as to the exact classification of a "sanitary engineer," whether it should

be "technical." "practical" or "scientific." But surely when we
consider the point, does not this difficulty of definition apply all
around? I cannot see that there can exist any such separate
:hing as technical education, no matter how much we may have
got used to the term, as applied to those forms of education which
are calculated to produce the man who actually performs. The
theory, or principle, of equilibrium has always appealed to me
as something of a natural force, producing the maximum efficiency. Therefore I do not ask, "Is a man technical?" "Is a man
scientific?" or "Is a man particularly this or particularly that?"
But I do want to know, are his actions based upon deduction or
are they merely instinctive? The principle of deductive thought
as separate from instinctive thought, appears to be the main
quality distinguishing the so-called " human " from the general
" animal " class of creation. Now, how can a man be " technical," as apart from " instinctive," unless his technical effective-
ness is based upon deductive formulae.

It is useful that a man should knock a nail skilfully, truly
and with precision, but if the action is controlled only by the
instinct which causes a bird to build its nest with precision, he
is no more than an instinctive animal. The man should, at the
back of each knock, feel and know all the reasons and accumula-
tive evidence which make for what is termed efficiency in work.
Absolute efficiency in production is found in machine results:
machine termination is the result of scientific application. The
man must be superior to the machine; and must, therefore, be
something more than a tool capable only of producing technical
results.

But what has this to do with sanitary engineering? Simply
this, you can no more make a purely efficient technical sanitary
engineer than you can make a purely efficient scientific sanitary
engineer. The efficient product must represnt a combination,
and must result in equilibrium. This fact is being recognized in
older countries. The United States of America in their universities, and European centres of research work as exemplified in
the Hamburg State Institute of Hygiene recognize sanitary cur-
riculums. Generally speaking, apart from the special sanitary
engineering course at McGill University, nothing is doing in
Canada. What is the result in Canada? We see, whenever an
expert is required to advise on sanitary matters, an outsider
called in. We train men in the problems of surveying and laying out a new country, in railway work, bridge building, and all
the requirements of transportation, in fact, in everything of a
utilitarian nature, even to the harnessing of our rivers for
power; but in the maintenance of ourselves in a normal state of
health, not only in health, but in a mere state of existence, we
have no curriculum of training.

I have used the term " a mere state of existence," because
one fact stands out plainly (all who run may read), a popula-

tion cannot grow beyond the available water supply at the rate of at least thirty gallons of potable water per head per day. Apart from our ideal and practical thought of dominion, our calculations of serial production, our schemes of railway development, grain elevators, and multitudinous efforts and methods of transportation and inhabitation, no two people can exist, and no community can grow beyond " available water supply."

Here comes in the " sanitary engineer." the man who can weigh up both physical and chemical data and draw deductions, and tell a community just how far it can grow, and where it must automatically stop. Now what does this mean from the sanitary engineering educational point of view? It means a man trained not only in taking levels, making surveys, measuring quantity and estimating the cost of work, but in estimating the amount of character of available water within a given area.

This takes in a great deal more than what is generally called "technical engineering." The character of any water, apart from its quantity, is surely most important. Underground water is not only water, but is also a mixture and chemical compound or compounds; its constitution depending upon the absorbed and soluble quantities of foreign matters taken up from strata. Surface water, in its constitution as far as admixture appertains, is dependent upon the absorption and mixtures of forms of vegetable growth and decay.

Water, as generally found, is, in fact, not water as represented by $H_2 O$. in our text books, but presents other and more intricate chemical formulae, sometimes inorganic, sometimes organic, generally a combination of both, but seldom water.

Now the conclusion of the above is simply this. Certain defined lines of education are required to include a curriculum of training which will produce not only a so-called " technical man," or man who can measure a quantity and show how the quantity can be delivered. but one who can also determine the exact quality of the quantity in its relation to public health. Such constitutes the true definition and training of a sanitary engineer in at least one of the aspects : I maintain that this definition holds good throughout. To put the matter in a nutshell. in a light plain to everyone, I would ask the question. " Where can I get the man trained in Canada as an engineer, who, if I put before him an analysis of a peculiar liquid or sewage, could produce on paper another analysis and show the cost and method of obtaining a defined change? I can get plenty of men from the schools who can do mechanical drafting or instrument work, but the man who can think and deduct from natural phenomena are few and far between.

Now, natural phenomena is not a hidden basis of action. but is the hypothesis of all formulae. Therefore, I consider that technical education should be based upon this ultimate basis of things. If I were asked to generally define the lines of education for a sanitary engineer, I would, by simply stating that there should be added to the general engineering course, a course in organic and inorganic chemistry, together with a knowledge of bacteriology as affecting the transmission of what are generally termed water borne diseases, as well as affecting the various fermentation processes for breaking up organic compounds.

There.are times when one has to listen to sentiments of depreciation applied to scientific men particularly, and to such phrases as " an ounce of practice is worth a ton of theory." All attempts to separate theory and practice are vain and foolish, and are the products of primitive ignorance. It is well to remember that the scientific man may be apt to know, while the other is apt to guess.

All that I have said up to the present may apply to the sanitary engineer, but I am inclined to think that the work of the Commission appertains to an enquiry more into the fitness or possible fitness of the man who may be employed by the engineer to execute or carry out his plans. The man who receives instructions and has to technically carry them out forms the material with which the Commission may have particularly to deal. This man cannot efficiently be produced by any method of any education apart from apprenticeship and practice in the particular line of work.

Sanitary engineering demands a host of so-called "technical men." The man who can actually go into a sewer trench and make a really watertight joint, or a lead joint in a water main, the efficiency of both sewerage and water supply depend almost entirely upon this man. In Great Britain we have associations of managers of waterworks and engineering installations which include so-called technical men, and who are simply the product of a demand. It strikes me that, after all, the law of demand and supply is the first cause instrument which produces the technical man, and his efficiency depends greatly upon the extent, character and amount of work which falls to his lot. The efficiency of the tool is in proportion to the efficiency demanded of the tool, and this simply brings me back in a circle to the point at which I started. The efficiency of technical work, if the product of deductive reasoning as apart from instinctive, depends upon the general and all-round training of the man who originates and who has the power to' direct completion.

A. H. Gregg, *of Wickson and Gregg, Architects, Toronto.*

I hope to see something done, not only for architectural students, but for the trades that come closely in contact with the people. To my mind the great difficulty is that the people lack interest in the trades they are following.

An example of this was brought to my attention recently. A printer in this city received a large order for books and found it necessary to engage nine additional printers. He had to discharge the nine men almost at once because he found that they lacked sufficient knowledge of type-setting. Now, it would seem a strange condition of affairs that these men actually lacked such an essential qualification of a printer.

A short time ago we were building a manufacturing plant in which we had to have a small power plant. One morning I went there and found that one of the plumbers had prepared a complete drawing of the principal apparatus, showing the location of every valve in it. This man had done so merely because he loved his work and had been taught to think.

Now it seems to me hopeless to try to teach men their trades unless they can be first taught to think. Their education should be brought to the point where they can reason out their work.

The printer I have mentioned told me that they could not get apprentices to stay, because their life was made miserable by the journeymen. In the profession of architecture we are in a peculiar position. For ages the architect had been trained very much as we read in "Martin Chuzzlewit," where Mr. Pecksniff, having received the pay for his instruction, turned the student loose to draw the cathedral, or what he pleased, allowing him to be his own master. Things have changed to-day. The larger offices are run on business-like methods. Few of the large offices, however, care to take on a man to teach him. They do not wish to be bothered with him. They do not think this the proper place for a student. It seems to me to be the business of the people to train architects, because there is nothing that comes more directly in touch with the citizen than the work of the architect. Evidently the government should take an interest in this question. It certainly is a point that should be brought to the attention of the Commission.

L. V. Redman, *Department of Electro-Chemistry, Faculty of Applied Science.*

Dr. Galbraith has pointed out that education could be theoretical sometimes, and sometimes practical; while under certain conditions it could be both theoretical and practical, but unless the practical could be well taught, it had better be left out. In this connection I might mention Professor Duncan, of the Kansas State University, who started a department in connection with that institution, the results of which have exceeded even the greatest ambitions of the founder, although the department is but four years old.

The idea is this: A graduate of the university enters the employ of some large manufacturing concern, in the field of research work. He continues using the laboratories of the university, and devotes a certain portion of his time to the problem they have set him. His other time is devoted to the teaching of undergraduates of the institution. A certain fixed sum is given him as a salary by the concern, and a bonus is also given him if he solves the problem upon which he has been working. One such man found the process of making what is known as salt-rising bread. He isolated the particular culture that was necessary to make the yeast. As the consumption of this class of bread reaches a value of approximately four hundred thousand dollars per year, and as the discovery was made at the expense of about fifteen hundred dollars, the results are indeed worthy of special note.

Another man worked on the caseins present in butter for the interests of those engaged in the shipping of large quantities of butter. He discovered a process which will net for his company the sum of one thousand dollars a day.

Hence it might be well to consider such a scheme in connection with our university. The man would be no expense to the university, except for the room he would require to work in, and the possible discoveries made by him would offset this. It seems to me that it would be a good investment, both for the university and any large company that employed such a man.

C. R. Young, *Department of Applied Mechanics, Faculty of Applied Science.*

Those in need of technical training may be divided into two classes: first, the professional class, composed of men destined by natural gifts or other advantages to occupy positions of large responsibility, and, second, the skilled laboring class.

From the fact that the representatives of the first class, viz., the practising engineers, or the engineer-managers, as they may be called, are occupied to a large extent in dealing with men, and as their power extends, to a less and less extent with things, the preparatory training for such should be essentially broad in its character. For this reason it is possible to train men to enter all branches of engineering at a single technical college.

The skilled workman is essentially a specialist, since proficiency in the handling of tools or in the carrying on of complex industrial processes is attained only after years of practice along a single line. It is even said in this connection that the difference can be noticed between a workman who has only his own experience behind him, and one whose ancestors for generations back were engaged at the same occupation. The skilled operator or workman, therefore, requires special instruction in the particular craft which he is to follow, and a technical school, to be of most value to him, must be most capable of affording assistance in his own branch of work. In this respect the training for skilled workmen differs essentially from that found most satisfactory for engineers.

A very important service which technical education may perform to assist the engineer in the conducting of his operations is to provide special instruction to those whose duty it is to carry his orders into effect. No one knows better than the engineer the difficulties encountered in avoiding errors of all orders of magnitude in the progress of engineering works. Sometimes this is due to carelessness, but most frequently to the inability of those engaged upon the work in the field to grasp the purpose and intention of the engineer. It may be due to lack of familiarity with plans and specifications, or inexperience in field operations, but the most serious errors arise from ignorance of the simple natural laws upon which the work is based. Thus, to cite actual instances, a contractor's foreman may place the vertical reinforcement of a retaining wall in the centre of the wall rather than at the back, or main reinforcement for a concrete beam at the top instead of at the bottom, and an inspector may quite as innocently decide that anchorages for the posts of a windmill tower are uncalled for, since the legs are battered, or that a form is capable of sustaining as much green concrete as can be dumped into it. As a means of lessening the likelihood of such errors as these, any arrangement by which contractors' foremen and inspectors on engineering or building work might receive elementary instruction in mechanics would be welcomed. At present the only means of this character generally available are afforded in the courses of the correspondence schools, but these can never equal personal instruction in efficiency.

T. R. Rosebrugh, *Professor of Electrical Engineering, Faculty of Applied Science.*

About twenty years ago I had brought to my attention a case which would be quite different in Toronto as regards the trade school. I refer to an institution in New Jersey, where there were but two industries engaged in—making of locomotives for the boys, and the silk industry for the girls.

It seems to me that the weakness of the present system is due to the separation of the two classes: the men brought up in the trades, and the men educated for professions at the universities.

I had some conversations some time ago with an engineer in charge of a large mechanical industry in this city, and he told me that it was next to impossible to get intelligent young men who would take an interest in some of the trades, for example, in foundry work. He thought it would be a great opening for any young man to devote himself to such work, and become a leader in that line.

Regarding the school system, it would appear that the child's time is considered as a by-product, something to be treated as having little value. If it were possible to include in the studies a few things that would be of vital importance to the young apprentice, at the same time allowing a good sound knowledge of the English language, no doubt much of the time spent in lengthy parsing of verbs could be devoted to instruction that would help the boy in the trade work of after years.

BOOK REVIEWS.

Heat **Engines** (being a new edition of *"Steam"*) by William Ripper; Longmans, Green & Co., 1909; cloth.

The character of this little work of some 312 pages is fairly well set forth in the preface where the author states that it "has been written as an introductory text book for the use of engineering students, and more especially for those who already have some practical acquaintance with the manufacture and use of machinery." It originally dealt only with steam machinery, but the present edition includes internal combustion engines as well.

The whole problem of the production and use of steam is dealt with in a very elementary way, beginning with the general effects and transfer of heat, the properties of steam, and passing on to the steam engine. The various parts and accessories of a steam engine, including the details of the machine itself such as the valve gear, the crank, cylinder, governor, etc., are then discussed along with the various types of condensers in use.

The compound engine is treated with a fair degree of completeness. One might have expected a somewhat more complete treatment of the indicator than has been accorded it, but as the book is intended to be introductory, probably no more space could be allotted for this purpose.

The chapters on the boiler and combustion are well arranged to give a knowledge of the conditions of combustion and economy.

The last two chapters deal with the steam turbine and the internal combustion engines, and the book concludes with a series of helpful questions bringing out the essential features of the various chapters. On the whole, the book is well illustrated, and should serve as a very good treatise for the beginner, and should also be helpful to the practical man. A number of problems are worked out through the book, and they are all put in such form that a knowledge of arithmetic is sufficient to understand them completely.

Reviewed by R. W. Angus, B.A., Sc., Professor of Mechanical Engineering.

C. M. Canniff, '88, has entered into partnership with Mr. John S. Fielding, as Fielding & Canniff, civil engineers, 15 Toronto Street, Toronto.

J. W. R. Taylor, '08, is in the employ of the William Hamilton Co., Peterboro'.

W. G. Swan, '05, is divisional engineer for the C.N.R., near Vancouver, B.C.

APPLIED SCIENCE

INCORPORATED WITH

Transactions of the University of Toronto Engineering Society

DEVOTED TO THE INTERESTS OF ENGINEERING, ARCHITECTURE
AND APPLIED CHEMISTRY AT THE UNIVERSITY OF TORONTO.

Published monthly during the College year by the University of Toronto Engineering Society

EDITORIAL

Applied Science begins it fourth year with the publication of this number. The new horse between the editorial shafts does not purpose making any material change in the direction of the journal's progressive course. Comparing it with his concept of the labor necessary to begin, three years ago, he finds it now an easy-turning vehicle. Rocks and ruts that rose prodigiously large then, are hidden now in dust.

Introduction '10-'11

For all this the graduates of the Faculty of Applied Science owe much to the former editor-in-chief, for the arduous effort and enthusiasm which the success of Applied Science during the past three years bespeaks. On behalf of them we wish Mr. Mackenzie at least the same measure of well-merited success in his new undertakings that they have seen the journal experience while under his supervision.

Since its debut, Applied Science has not failed as a link of communication and a binding tie between the undergraduates and the alumni. From its pages the latter have read the papers delivered to the former as members of the Engineering Society. The benefit derived thus has been mutual.

Co-operation

The journal has, in turn, profited by the loyal co-operation the alumni has given it. The result is a recognized leader in the broadcasting of technical and engineering literature.

Opportunities for advancement are many, and a voice from its readers throughout its expansive field in Canada and the United States, in the form of papers, discussions, or plain, unvarnished criticisms will aid very materially in this advancement.

Every issue contains an article of more or less especial interest to every graduate; a problem, perhaps, concerning which his own candid opinion has, for some time, sought an airing: a subject upon which a paper may have been commenced, but never finished, or finished but never submitted. Let us hear from you.

One of the articles in the make-up of this issue is a synopsis of the discussion on the problem of technical education, which took place at a meeting of the Engineers' Club Club on the evening of Oct. 13th. It is evident, from a perusal of the article, as well as from the study of general industrial conditions, that the Commission on Technical Education has before it in its endeavor to produce an improvement along educational and industrial lines, a task as gigantic as the freedom of scope and powers of investigation it possesses. Although the engineering profession is but one of the numerous branches subject to its examination, it is, at all events one of the most important, and as some will say, one of the most needful of improvement. It may be expected, therefore, that the Commission on Technical Education will enact some very helpful suggestions as a result of its thorough investigation.

Vacation
Work

One problem there is, however, for which we need scarcely expect from them a speedy solution, and which, consequently, might well be considered by ourselves, namely, that relating to vacation work for under-graduates. Every October returns to us men who have spent the summer in search for experience they know to be necessary for the rounding out of their college training. Every October hears various opinions as to the efficiency of this vacation work in connection with the theoretical training received during the college year. The reasonableness of the variety of opinions is well worth considering. Would an increase of thought and consideration as to a more satisfactory combination of the electrical course or the mechanical course, and the vacation work the student obtains be worth the trouble? Obviously so, as this is the only present outlet into the broader channels of co-operation between the course in, say, mechanical engineering, and the mechanical industry itself. The engineer-to-be must combine theory with practice somewhere. The complete undertaking is impracticable in the college alone, because mechanical equipment and methods of mechanical operation are too changeable to be kept up-to-date, since their speed curves do not coincide with that of the financial resources of the college.

If we step thirty years backward we note the establishment of electrical engineering courses as a response to the call of the industry itself. The same applies to the mechanical and other engineering courses—each the result of demand. The technical course is, in every case, the servant to the practical industry.

Methods are being employed in Great Britain, in Germany, and in the United States, by engineering institutions, to develop as high a degree of efficiency as possible in the service they render to industrial development. They do well to utilize the voluntary assistance of the manufacturers themselves, since neither body can alone substantially lessen the gap between, let us say, the ideas occupying a student's professional mind upon leaving the factory in September, and those which slowly trickle in after a few weeks in the lecture room. (The fact that it requires several weeks at both ends of the college year to adapt one's self to one line of mental activity after having pursued another for some months, is in itself an indication of a greater diversion between them than should exist). Both the employer and the university are trying to bring into alignment these separate occupants of the student brain. What they might do working in conjunction ought to be a topic interesting to both, and would undoubtedly prove interesting to the under-graduate. as it would be the means of moulding college course and apprenticeship course into a more compact form, perhaps lessening by a year or more the time required at present to complete both.

The shortest distance between the two points of inexperience and industrial efficiency is along the line of co-operative education. In view of this, a commission consisting of representatives of, let us suggest, various firms in the electrical industry, and of men from the departmental staff of the engineering school could, without making gigantic concessions on either side, map out a course of training that would render the students' vacation work of more assistance to his electrical engineering course, and thus of more assistance to his industrial life.

The other departmental courses could be materially improved by much the same procedure.

OUR REPRESENTATION IN THE SENATE.

The Result of October Elections.

The elections of a month ago resulted in the re-election of C. H. Mitchell, C.E., and E. A. James, B.A. Sc. to represent the graduates of the Faculty of Applied Science and Engineering in the senate chamber of the University. Both these men have taken at all times a very deep interest in the affairs of the Faculty of Applied Science, and in the progress in engineering that its graduates are affecting in the Dominion and elsewhere. Both are well qualified, from an engineering point of view, for the work that has been entrusted to them by these graduates.

E. A. James, B.A. Sc.

Mr. Mitchell, and his experience as a practising engineer, are well known throughout Ontario, especially in the Niagara Falls district, where he spent his first eight years after graduating, in designing and managing the construction of various municipal projects, chiefly Hydro-Electric. Later he became chief of the Mechanical Department of the Ontario Power Company, of Niagara Falls, which position he retained until 1905. Since that date Mr. Mitchell has been a resident of Toronto, a member of the consulting firm of C. H. and P. H. Mitchell.

Some years ago Mr. Mitchell made a tour of inspection through Europe, investigating the results there of modern engineering practice.

He has been for some time a member of the executive of the Canadian Society of Civil Engineers, and two years ago was chairman of its Toronto branch.

Mr. E. A. James is a more recent graduate of the Faculty of Applied Science, and consequently, is more in touch perhaps, with the majority of younger graduates. He was a member of the class of '04, and in the following year was elected by the Engineering Society as its president. Later he became associated with several civil engineering enterprises in western Canada and in northern Ontario, where he was superintendent of ciated with several civil engineering enterprises in western Canadian Pacific Railway.

Some 3 years ago Mr. James entered the field of journalism, and as managing editor of the *Canadian Engineer*, has evidenced his ability by the journal's rapid development into the service it now renders to the profession in Canada.

Mr. James is still actively engaged in engineering himself, being town engineer for North Toronto. He is an Associate Member of the Canadian Society of Civil Engineers.

In all there were five candidates for the two senatorial chairs. These were C. H. Mitchell, '92' of the firm of C. H. & P. H. Mitchell; A. F. Macallum, '93' engineer for the city of Hamilton; W. E. H. Carter, '98' of E. T. Carter & Co.; E. A. James, '04, of the *Canadian Engineer*, and T. H. Hogg, of the Ontario Power Company, Niagara Falls. To the successful candidates Applied Science extends sincere and hearty congratulations.

C. H. Mitchell, C.E.

BIOGRAPHY.

R. W. Leonard, C.E.

Reuben Wells Leonard, the contributor of the first article in this number of Applied Science, is a native of Brantford, Ont., and a graduate of the Royal Military College, Kingston, winning a silver medal in the class of '83.

Since graduating he has followed his profession for the most part in Canada, but for some little time in the United States and in Central America, taking up railway construction, mining, and water power engineering.

For two years following graduation, Mr. Leonard was instrument man and engineer in charge of construction on the Canadian Pacific Railway north of Lake Superior. In 1885 he served in the North-West Rebellion on various staff appointments. Later he acted as engineer on Canadian Pacific Railway surveys and construction work in Manitoba and Ontario.

From 1886 to 1890 he was chief engineer for the Cumberland Railway and Coal Company, of Nova Scotia. For the following sixteen years Mr. Leonard has served as engineer and manager of construction on various portions of the Canadian Pacific Railway from Quebec to British Columbia, on the St. Lawrence and Adirondack Railway, on the Rutland Canadian Railway, on the Cape Breton Railway, and as contractor on the Parry Sound Railway.

In water-power engineering he has constructed hydroelectric plants at Niagara Falls, Ont., near St. Catharines, Ont., and for the Kaministiqua Power Company at Kakabeka Falls, Ontario.

In 1906 the Coniagas Mines, in which Mr. Leonard has a controlling interest, created no small stir in the mining world, and carried away much of his attention from his civil engineering projects. About this time he built a smelter at Thorold for the refining of silver ores. In 1909 he became actively interested in the copper claims at Bruce Mines, Ont., in which work is being carried on briskly at present.

Mr. Leonard is president and general manager of the Coniagas Mines, Limited, with its subsidiary enterprises, the Coniagas Reduction Company, at Thorold, and the Redington Rock Drill Company. He is likewise president of the Bruce Mines, Limited. He is vice-president of the Canadian Society of Civil Engineers, and vice-president of the Canadian Mining Institute.

Last spring Mr. Leonard received an appointment to the Board of Governors of the University of Toronto.

Herbert Johnston, '03' is town engineer for Berlin, Ont.

A. A. Ridler, '07' is with the Constructing and Paving Co., Toronto.

ENGINEERING SOCIETY.

As in previous years, the Society executive is alternating its general meetings with others for the different departments. This plan has met with much success, since it was inaugurated, in providing papers and discussions that sustain the interest of all the members.

The first general meeting for the year was held on Wednesday, Oct. 5th. It was addressed by President Falconer, Dean Galbraith, and R. W. Leonard, C.E., and was attended by quite a number of the graduates, in addition to a full attendance of student members.

The resignation of W. A. Gordon, '10, newly-appointed treasurer for the Society, was accepted, and the ensuing nomination resulted in the election by acclamation, of M. B. Gordon, '10.

On Oct. 19th, the various sectional meetings of the Society were held. The electrical and mechanical sections heard Mr. H. W. Price; the civil and architectural sections, Mr. T. R. Loudon, while the chemists and miners were addressed by H. L. Batten, '11.

On Wednesday, Nov. 2nd, Mr. T. Aird Murray, C.E., spoke to those present at the general meeting on "Water Purification." His interesting and instructive paper was well illustrated by slides.

On Monday, Oct. 24th the men of the fourth year went to Niagara Falls, and investigated the various hydro-electric installations there. The excursion was in charge of Professor Angus and Mr. Price, and was taken advantage of by over one hundred men.

WHAT OUR GRADUATES ARE DOING.

This section is conducted with a double object in view: first, to give the graduate professional news of each other; second, to give the undergraduates an idea of the possible fields of employment open to them in the future.

W. A. Gordon, '10, is in business at Sundridge, Ontario, as a member of the firm of Gordon & Thornton, lumber merchants.

Guy Morton, '09, and R. S. Davis, '07, are managing the Calgary branch of the Canadian Westinghouse Co.

C. J. Porter, '09, is with the B. C. Electric Railway Co., Vancouver.

Gordon Kribs, '05, is with Smith, Kerry & Chace, as manager of their newly-opened branch in Portland, Oregon.

A. A. Kinghorn, '07, is in the Roadways Department, City Hall, Toronto.

W. D. Black, '09, is with the Otis-Fensom Elevator Co., as superintendent of their eastern branch, with headquarters in Montreal.

E. A. Jamieson, '10, is C.P.R. inspecting engineer for the Okanagan district, Kamloops, B.C.

T. A. McElhanney, '10, is on geological government work in Vancouver.

A Gillies, '07, is resident engineer on the construction of a power plant at Minnedosa, Man.

A. J. McPherson, '93, has been appointed municipal commissioner for Regina, Sask.

E. L. Cousins, '06, is assistant engineer for the city of Toronto.

W. G. Turnbull, '09, and P. J. McQuaig, '09, are in Milwaukee, Wis., with the Allis, Chalmers Co.

Leslie R. Thomson, '05, has been appointed lecturer in drawing, University of Manitoba.

F. F. Wilson, '09, is with a Mitchell survey party at Smoky Lake, Alberta.

George A. Tipper, '09, is also on survey work, with A. L. McNaughton, D.L.S., at Heatherwood, Alberta.

H. W. Fairlie, '10, has accepted an appointment with the Tungstolier Company, of Canada, Limited.

Wilfred C. Cole, '10, is with the Central Colorado Power Company, in their Shoshone hydro-electric plant, on the Grand River, Colorado. This company is supplying the city of Denver with the major part of its lighting and power requirements by a transmission line 150 miles in length, over the Rockies. The power is supplied at 100,000 volts.

G. E. Woodley, '10, has since graduating, been with the Westinghouse Electric and Mfg. Co., East Pittsburg.

C. R. McCollum, '09, and W. J. McIntosh, '09, are in the employ of the Otis-Fensom Elevator Co.

C. E. Toms, '09, is with a survey party in British Columbia.

G. E. McLennan, '10, and A. S. McArthur, '09, are on D. L.S. work near Prince Albert.

G. E. D. Greene, '09, is on railway survey work in eastern Ontario.

Othmer Ross, '10, is in the draughting office of the Dominion Bridge Co., Lachine.

Among the recent staff appointments in the Faculty of Applied Science and Engineering are the following:

H. E. T. Haultain, C.E., has been appointed Professor of Mining.

J. J. Traill, '05, and W. W. Gray, '04, have been appointed lecturers in mechanical engineering.

J. A. Stiles, '07, and R. E. W. Hagarty, '07, are demonstrators in drawing.

W. C. Blackwood, '06, is demonstrator in the department of physics.

N. H. Manning, '09, has been appointed demonstrator in thermo dynamics.

H. A. Cooch, '09, is demonstrator in the department of electrical engineering.

J. T. Lagergren, formerly of Sweden, has been appointed lecturer in machine design, and is in charge of the draughting room work of that department.

J. E. Keppy, '06, has accepted a position on the staff of the Canadian Inspection Bureau.

A. U. Sanderson, '09, is at present at work on the new filtration plant.

OBITUARY.

It is with sorrow that we report to our readers the death of H. S. Fierheller, B.A. Sc., a member of the class of '06, which took place at his home in Toronto in May, 1910.

Index to Advertisers

Applied Science

INCORPORATED WITH

TRANSACTIONS OF THE UNIVERSITY OF TORONTO ENGINEERING SOCIETY

Old Series Vol. 28   DECEMBER, 1910   New Series Vol. IV. No. 2

THE PURIFICATION OF PUBLIC WATER SUPPLIES.

T. AIRD MURRAY, C.E., M. Can. Soc. C. E.

Professor Hyde in a recent report upon a water supply for Sacramento, assumed that the question of quantity was of much more importance than the question of quality in the first instance. It is taken for granted that, no matter what the quality may be in the raw state, as long as the quantity is sufficient to meet all future wants, the eventual quality can be guaranteed by the adoption of purification methods.

Now this assumption as a basis for a report on a probable water supply has only been made possible of late years. Custom was at one time to insist upon an analysis of any probable water supply, and on the strength of that analysis to either accept or disregard it. This custom, however, is now subject to certain qualifications.

We know that no community can grow beyond its available pure water supply. Pure water is an absolute necessity to a community. If we have an island containing no potable water, no man can exist on that island. If there is only water for one man, then one man only can exist, and so on. No matter what latent wealth that island may contain, without potable water or the means of delivering potable water, that wealth is useless.

So in Canada where we have great areas of land wealthy almost beyond measure in productive fertility, the realization of such wealth is subject to the supply of potable water. There are town sites on the prairie which can never become practically more than town sites; there are villages which cannot become towns; towns which cannot become cities, and cities which cannot grow beyond a limited population, because of the fixed quantity of available pure water.

At the present day the amount of available pure water is much greater than it was, say, ten years ago; because we now know of efficient means by which impurities either inorganic or organic can be removed from water, and so water which at one time would, on analysis, have remained condemned may

now be classed as potable. This is the meaning of Professor
Hyde's assumption. We can now go straight for quantity
leaving the question of quality to be taken care of as an after-
thought. This does not mean that quality has no bearing on
the choice in the first instance. It only means, as it did at
Sacramento, that although a pure ground water supply existed
in territory to the south east of the city estimated at 20 million
gallons per day, that Hyde advised the city to pump from a
polluted river source with an unlimited supply which would
meet the probable growth of the city. In fact, the original pure
water supply would have limited the city growth, whereas the
abundant originally impure supply satisfied an unlimited city
growth. A few years ago the growth of Sacramento was
limited, by means of our newer knowledge of treating impure
waters, its growth is now unlimited.

As to the modern requirements of a pure water supply,
allow me to quote from Hyde's report above referred to:—

(a)   The supply must be abundant and unfailing.
(b)   The water must be free from pathogenic
germs.
(c)   The water must be free from those allied
organic forms which may not as yet be recognized as accompanying disease, but which
may, nevertheless, not be conducive to health.
(d)   The water must not be discolored.
(e)   The water must at all times be free from
taste and odors.
(f)   The water must be uniformly clear and free
from turbidity, both that which may be produced by suspended mineral matters and
that which may be due to suspended organic
impurities.

To the above may also be added that water should not contain
more than 15 degrees of total hardness and 5 of permanent.

It is very difficult and rare to obtain water which meets all
of the above requirements. Settled large basins of water
present the nearest approach to the ideal. Almost all rivers,
whether pathogenically polluted or otherwise, are apt to
present turbidity at times. Spring waters, though free from
turbidity and generally pathogenically pure, are apt to contain
the salts peculiar to the strata in which they are found, and
are hence generally excessively hard, and at times even unfit
for domestic use unless artificially softened.

It will be impossible for me in the limited time at my
disposal to treat of methods for removal of all forms of impurities from water. I intend, therefore. to confine my remarks
more particularly to methods and principles applied chiefly for
the removal of pathogenity.

Filtration has been and remains the most generally
accepted method of purifying water. We know that water

in passing through the surface layers of the earth very quickly loses all trace of original organic impurities, and the only impurities it may contain must be new ones picked up by contact with the earth. It is, therefore, apparent that if we can artificially reproduce and improve upon the best conditions which exist in earth surface layers, we shall produce a method of water purification. That method has been evolved in the system called "Slow Sand Filtration." There is nothing of a mysterious nature, nothing of a complicated chemical character, takes place with slow sand filtration. The whole process simply amounts to a method of straining. There are certain engineering features connected with the regulation of the water as applied; but beyond these, the filter is simply a bed of sand of a given area and thickness capable of allowing a given quantity of water to percolate from the surface to the under-drains. The filters are subject to constant saturation, and are worked with a depth of water covering the whole surface sufficient to overcome the friction produced by the sand. The quantity of water which can be dealt with daily varies from 1¾ million gallons to 6¾ per acre, depending upon the quality and character of the raw water and the size of the sand grains used.

We have said that the whole process is simply one of straining. Thus, it differs almost entirely from the sewage filter, the object of which is not only to strain or hold back, but also to bring about certain chemical changes in the filtrate. Sand filtration does its most important work on and in the surface layer to a depth of about one quarter of an inch. The sand pores are not sufficiently fine in themselves to effectually keep back bacteria. The first few days' working of a sand filter is practically useless as a means of removal of pathogenic impurities; but as a scum of matter, deposited from the water, forms a carpet over the whole surface, bacterial removal efficiency increases to almost 100%. This scum formation is the peculiar feature which gives to slow sand filtration its unassailable position as the best mechanical method of removing pathogenity from the water. Like all human contrivances, however, it has its drawbacks. Chief of these is, that the surface scum continues to thicken until it becomes almost impervious and must be removed. Covered in sand filters average in cost on this continent from fifty to sixty thousand dollars an acre and to this must be added cost of pumps, pipe connections and all the other necessary appurtenances. The cost of operation ranges from two dollars per million gallons at Mt. Vernon, N.Y., to almost six dollars at Lawrence, Mass.

Slow sand filtration has been termed the English system of treating water. It has been adopted largely in Europe and latterly in the United States, and is now being installed in connection with the water supply of the City of Toronto.

It may here be said that the introduction of water filtration has universally been followed by a marked reduction in the so called water borne diseases, especially in the typhoid rate. At Hamburg the typhoid death rate averaged 47 per 100,000 for the five years previous to the installation of filters, and only 7 for the five years following. At Zurich, before the introduction of filters the typhoid death rate was 69 and after filtration 10. At Lawrence, Mass., the figures were 113 and 25 under similar circumstances. At the present time over 30,000,000 people throughout the world are being supplied with filtered water, more than one third of this number being in Great Britain. London, Liverpool, Birmingham, Leeds, Sheffield, Dublin, Leicester, Newcastle and Edinburgh are among some of the large cities in Great Britain using sand filters. In Europe we have Hamburg, Berlin, Breslau, Chennitz, Madenburg, Altona, St. Petersburg, Warsaw and Antwerp, where such filters have been in successful operation for many years. There are also similar plants at Calcutta, Bombay, Agra, Shanghai, Hong Kong, Tokyo, Yokohama, Osako and other cities. In the United States there are slow sand filters at Philadelphia, Pittsburg, Washington, Albany and many other places. In Canada there is one plant at Victoria, B. C. The typhoid death rate for Canada is 35 per 100,000, for European countries 16.

Slow sand filtration is an effective method of purifying water. It is costly both in construction and operating expenses. It requires exact and careful management; this applies to all methods of water purification. A bacteriological laboratory should be run in connection with the plant. It is in every sense more suited to wealthy and large communities than to small.

There is another form of filtration more distinctly American called "mechanical" or "rapid filtration." This system at once appeals to the layman because of the small space it occupies. Whereas slow sand filters will treat on the average 2½ million gallons per day, rapid filters average 120 million gallons per day. The sand used is coarse grained, and efficiency in purification depends not so much upon the filter itself, as upon preliminary sedimentation accompanied by the use of a coagulant. The filters are generally installed in units equal to about a hundreth part of an acre. High purification efficiencies are obtained by technical care in regard to proper working and the scientific application of the coagulant. Generally speaking the bacterial removal efficiency is below that of slow sand filters. The makers, and there are many, do not as a rule guarantee a removal of more than 97% of total bacteria, whereas 99%, and over, removals are looked for with slow sand filters. The temptation to use mechanical filters in this country is very great, as they can be installed easily in a building in connection with the pump plant and so guarded from severe frost.

With reference to the use of a coagulant I will quote from

a report of the "Joint Special Committee to examine and report relative to the pollution of Water Supply, and the best method of filtration", City Document No. 15 of the city of Providence, R. I., as follows:—

"If the diameter of matter floating about in water is much less than that of the interstices between the grains of sand composing the filter bed, such matter, except as much as is caught upon the sharp edges of the quartz, will go right through the filter with the water.

"Now, if a substance could be introduced, drop by drop, into the water before it comes to the filter bed, which would have the effect of curdling the matter together, so that every one hundred or so of the smaller particles were made to join together and become one large particle, much as vapor or steam is condensed into drops, it would follow that they would be caught and held from going through the filter. This is accomplished by adding alumina (alum) to the water as it flows to the filter.

"The amount required is from almost none at all to about three quarters of a grain, according to the state of the water, say, an average of from one quarter to one half grain per gallon in the ordinary condition of the Pawtucket River water.

"The action is the same as when coffee is cleaned by means of the white of egg. No white of the egg goes to the drinker of the coffee—it is all drained off with the grounds; and as no alum goes to the drinker of the water, it unites with the impurities in the water and settles in feathery flakes of insoluble hydrate on the top of the filter, and is washed out with its accumulation of impurities when the filter is cleaned.

"The analysis of the purified water shows no trace of the alumina used, while the analysis of the wash water shows that the alumina is all washed out with other impurities. This feathery bed of precipitate flakes produced by the alum forms a filtering media of insoluble mineral matter which is well nigh perfect in its character. Bacteria are like the very fine particles of clay of some water, so small as to pass the sand or quartz, but they are caught by the feathery precipitate of alumina hydrate, much as the bacteria contained in air are prevented from entering a phial closed with sterilized wool."

So we see that, first, as in the case of the slow sand filter. so with the mechanical, bacterial removal efficiency does not so much depend upon the filtering material itself as upon an artificial surface or blanket being formed.

One of the most successful installations of mechanical filtration is that at Harrisburg, Pa. The average bacterial removal is above 99%, being 99.62% for the year 1908.

Now the assertion may be made that these methods of filtering water are not perfect. Efficiencies are given of something over 99%, but what about the remaining bacteria, the percentage number of which must depend upon the original number to start with. For instance there were days last year when the Harrisburg water presented as many as 85,250 bacteria colonies per c.c. Now a 99 per cent. removal would yet leave over 800 bacteria per c.c. in the water.

The answer to the above assertion is contained in the one word "sterilization." If a water is so impure from a pathogenic point of view that it requires to be filtered and it is necessary to guarantee it at all times then "disinfection", or "sterilization" as it is sometimes called, must form a final adjunct to any filtration process. This principle has been recognized for some time in Europe as illustrated by the number of ozone treatment plants which have been added to filtration plants. In many of the European rivers the bacterial counts run so high, that it is impossible to obtain by any method of filtration results which come anyway near the recognized standard for filtered water, viz, not more than 100 bacteria per c.c.

The principle is also being recognized in this continent. Nashville, Minneapolis, Quincy, New Jersey and many other towns, even including Harrisburg, have adopted methods of disinfecting the water, while in Canada we have several temporary plants installed in Toronto, Montreal, Pembroke and several in Western Canada.

Disinfection by ozone has not as yet found much favor in this continent, the disinfectant chiefly used being chlorine obtained from chloride of lime. With the Siemens, Halske & De. Fries system, efficient disinfecting results can be obtained with ozone. The process, however, of forcing contact between the ozone and the water is expensive, and even in cases as at St. Maur, Paris, costs more than the production of the gas. The difficulty arises owing to the fact that ozone is practically an insoluble gas in water, hence the difficulty of obtaining contact with every drop of water. In the case of chlorine in the form of hypochlorite we have a salt which is soluble in water, and which if properly mixed reaches every particle of water. The well known disinfecting action of chlorine depends upon its power of combining with the hydrogen of water and liberating the oxygen; thus $CL_2 + H_2O = 2 HCL + O$.

This atom of oxygen is said to be nascent at the time of liberation and in this form acts as a most powerful disinfectant. The bleaching power of chloride of lime is due to the same reaction. As long as the lime is kept dry it has no bleaching action, but when in contact with moisture the reaction at once takes place.

The high disinfecting efficiencies obtained by mixing chlorine with water has, of late, brought it into great favor as a temporary expedient in cases of typhoid outbreaks due to impure water. The plant necessary for treatment can be fixed up practically in one day. The cost is a mere bagatelle, and the operating expenses run about 50 cents per 1,000,000 gallons of water treated.

The plant need not consist of any more elaborate detail than two or three 60 gallon casks with a small tank supplied with a ball cock to maintain a constant head fitted with a regulating tap at the orifice.

Hypochlorite is most economically and readily obtained from chloride of lime, the lime containing on the average about 33 per cent. of available chlorine. It is usual to mix in the barrels a one per cent. solution of the lime with water, viz., at the rate of 1 lb. of chloride of lime to 100 lbs. or 10 gallons of water. The lime settles out as calcium hydrate, and the liquid is led to the orifice tank from which it is dosed into the water supply in the necessary proportion.

The proportions of available chlorine to the amount of water requiring disinfection varies with the character of the water. Turbid water and water containing considerable amounts of vegetable matter require much more chlorine than waters which are practically free from these conditions. Ontario Lake water as normally represented by the quality at the intake is efficiently disinfected by the addition of .33 parts of available chlorine to 1,000,000 parts of water. Some waters require not more than .15, while others will practically eat up chlorine before it has any chance to act as a germicide. The reason for this variation is the affinity of organic matter to oxygen. No hard and fast rule can, therefore, be laid down for its application. Where a water may vary in condition. relative to turbidity and organic content, so the amount of chlorine must also be varied. The application is not dependent upon any hard and fast rule, but must be based upon a scientific knowledge of degrees of cause and effect.

A water which can be most efficiently treated with any disinfectant which relies upon nascent oxygen, should present constant conditions. If these constant conditions are not peculiar to a raw water, then they should be assured by some preliminary treatment before the disinfectant is applied. The preliminary treatment may take the form of sedimentation, slow sand filtration, mechanical filtration, or a combination of sedimentation and filtration, depending upon local conditions. This practically means that in the majority of cases disinfection is not likely to supersede the accepted general methods of purifying water, but that it will merely form an adjunct or final process, by which any water can be absolutely guaranteed as free from disease germs.

The introduction of disinfecting processes will probably in the future tend to a more favorable acceptance of rapid filter methods rather than slow methods. The main reason as we have seen, for a preference for slow sand filtration as against mechanical filtration, has existed in the higher efficiency of the former as a germ remover. Our newer knowledge that water of a constant character, as turned out by a mechanical filter, can be sterilized at a small cost, has caused many of the advocates of slow sand filtration to reconsider their former conclusions.

The question has been asked. whether the addition of hypo-chlorite deleteriously affects water for purposes of domestic use.   The answer is, that the small amounts of chlorine required, do not in any way affect the water.   Where it might be necessary to use large amounts of chlorine, the water must of a necessity receive preliminary treatment to allow of only reasonable amounts being used.   The author is of opinion that our present knowledge of the subject limits the permissible amount of chlorine to about .50 parts in 1,000,000.

At the above rate of application it would require with chloride of lime containing 33 per cent. of available chlorine, as follows:

  1.5 lbs. of chloride of lime to 1,000,000 lbs. of water.

  =1.5 lbs. of chloride of lime to 100,000 gals. of water.

  =15 lbs. of chloride of lime to 1,000,000 gals. of water.

In order to make a 1 per cent. solution sufficient for 1,000,000 gallons of water. it would require the 15 lbs. of chloride of lime to be mixed with 1500 lbs. of water, or 15 lbs. of lime to 150 gallons of water.

There is no doubt that disinfecting processes of water treatment have come to stay, just what form their exact rela-tion will take up relative to older methods being yet somewhat undecided. However, I have as far as possible endeavored to lay before you the present situation relative to purification of water as far as the removal of pathogenity is concerned. Time will not allow of entering into the hundred and one other impurities which it may at times be necessary to treat. We have algae growths, vegetable iron growths, such as ferrigmous slime. spongidae, all kinds of complications caused by inorganic matters in solution, all of which are interesting and capable of remedy.

Sufficient is it. if I have shown you that waters, which at one time would have remained condemned as possible sources of water supply, may now be brought into general use.

MERCURY ARC RECTIFIERS.

H. A. COOCH, B.A.Sc.

The mercury arc rectifier was first introduced to meet the demand for some device for rectifying from alternating to direct current, economically, efficiently, and without occupying too much space in the power station. It is the case in a great many stations and substations at present that only alternating current machines are in use. As a result, it would necessitate a great outlay to install direct current machinery to supply d.c. needs. In such cases it is apparent that this arc rectifier is to play an important part in the betterment of local substation economics.

The first to realize the rectifying properties and possibilities of mercury was Mr. Peter Cooper-Hewitt, of the Cooper-Hewitt Mercury Arc Lamp Co., now absorbed by the Westinghouse interests. In his investigations he realized that mercury was of such a nature as to permit of rectification from a.c. to d.c. In April 1900, he perfected his first rectifier, the patents for which were secured in Jan. 1902.

At present there are some half dozen methods of obtaining direct from alternating current, all of which have disadvantages connected with them. Many of these disadvantages practically disappear, however, in the case of the mercury rectifier.

In the first place the motor-generator set is quite frequently used for the a.c. to d.c. transformation. This however is high in first cost and requires large floor space for installation. The efficiency at full load of sets of proper size for charging vehicle batteries has been comparatively low, and at light loads very low. In the case of high voltage d.c. for constant current series arc lights it would be impossible to obtain proper commutation on the d.c. machine, hence the operation of a number of series lights from a motor-generator set is an impossibility.

The single phase rotary converter is another method of rectification. However, it is not as flexible as the motor-generator set, particularly as regards voltage, and it requires more care and higher intelligence in starting and operating it. Besides, as in the case of the motor-generator set, we cannot commutate a sufficiently high d.c. voltage to maintain a set of series arc lamps.

Together with the above two most important methods we have a number of others which have proved more or less unsatisfactory under certain conditions. These include The Synchronous or Mechanically Driven Rectifier, sometimes called the Lethuele Rectifier, the Chemical Rectifier, in which aluminum is the important substance, due to its peculiar polar resistance; and Switching Devices for Rectification. There are also various forms of Electrolytic Rectifiers involving substances such as tantalum, carborundum, etc.

There are two distinct types of mercury rectifier, namely, low tension and high tension types. The storage battery charging outfit is a typical example of the first form where the pressures run below 300 volts.

The low tension set consists essentially of:

    (a)   An auto-transformer for excitation.
    (b)   A direct current reactance.
    (c)   The rectifier bulb and holder.
    (d)   An operative panel.
    (e)   A regulating reactance.

The auto-transformer, or compensator reactance, as it is sometimes called, is simply used to step the line voltage up or down a sufficient amount to give the desired terminal a.c. volts across

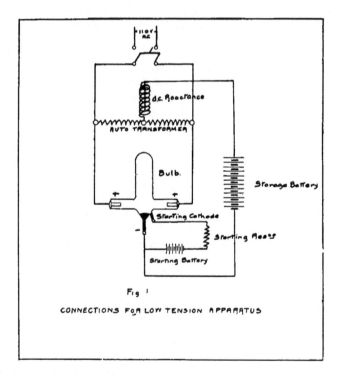

Fig 1

CONNECTIONS FOR LOW TENSION APPARATUS

the bulb. It consists in principle of two coils of an equal number of turns connected in series with one another, the middle point being connected through the d.c. load to the cathode. This is shown in Fig. 1.

In order to smooth out the otherwise rapidly fluctuating waves and also to cause the instantaneous current in one anode or cathode circuit to overlap the instantaneous current in the other by a small angle usually about 20°, we have sometimes a d.c. reactance in series with the d.c. load.

The Bulb and Holder.

The bulb or tube is a glass vessel exhausted to a high degree and sealed at the tip, the vacuum being necessary for the unimpeded flow of current. It is equipped with two anodes, one cathode, and one starting anode. The cathode and starting anode are filled with mercury. The production of the conducting vapor takes place at the surface of the cathode from what is known as the cathode spot. This cathode spot or bubble continually wanders around on the surface of the mercury cathode.

The size and shape of the bulb depends almost entirely on its capacity in amperes and volts, since it can be easily seen that higher voltage types require more protection against spatterings of mercury, and the consequent injurious effects from this cause, than the low voltage types; while the cross sectional area depends on the current required.

The rectifier bulb is supported at the back of the board by a holder consisting of a metal frame so pivoted at its centre that it can be rocked back and forth. The life of the bulb under normal operating conditions is at least 400 hours unless defective in manufacture. In fact the life of the bulb is almost a matter of guesswork unless perhaps it refuses to operate at all. Tests so far have shown an average of

3000 hours for the 10 ampere tubes.
2500 hours for the 20 ampere tubes.
700 hours for the 30 ampere tubes.

On the operating panel are mounted usually voltmeter, ammeter, double throw switches for connection to the supply and load circuit, and the necessary double pole and single pole switches for starting and operating the rectifier. Fuses and circuit breakers are provided for protecting the bulb against overloads. A starting resistance is mounted on one of the pipe supports and is connected in multiple with the pilot lamp mounted on the front of the board. The pilot lamp is used to indicate that the rectifier is in operation, and also acts as a warning that the starting resistance switch should be opened as the load is thrown on, as the lamp is dark when the rectifier is operating on the load only. The necessary reactance dials are also provided. The connections behind the panel can be seen in Fig. 2.

A smoother or finer control of d.c. voltage as well as current is obtained by a regulating device in series with the a.c. supply line and mounted on the pipe support back of the panel. It consists of a coil wound on a laminated iron core, a number of taps are taken from this coil and connected to a semi-dial switch mounted on the panel.

The high tension series arc lighting type consists of the following apparatus:

1. A constant current transformer.

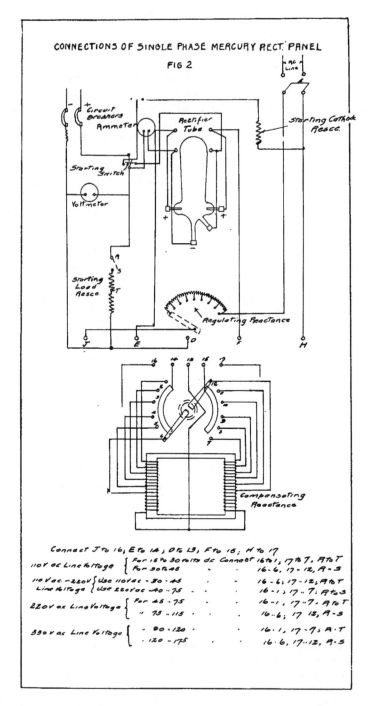

CONNECTIONS OF SINGLE PHASE MERCURY RECT. PANEL

FIG 2

Connect J to 16; E to 14; D to 13; F to 16; H to 17

110 V ac Line Voltage { For 16 to 30 volts dc Connect 16 to 1, 17 to 7, A to T
For 30 to 48   "   16-6, 17-12, A-5

110 Vac -220V { Use 110 vac - 30 - 45   "   "   16-6; 17-12; A to T
Line Voltage { Use 220 vac 40 - 75 ..   "   "   16-1, 17-7; A to 3

220V ac Line Voltage { For 45 - 75   "   "   16-1, 17-7. A to T
   " 75 .. 115 .   "   16-6; 17 13, A-5

330 V ac Line Voltage { " 90 -120 ·   "   16-1, 17-7; A·T
· 120 - 175   "   16-6, 17-12, A-5

Plate 2

2.   A d.c. reactance.

3.   An exciting transformer for starting.

4.   The rectifier bulb.

The transformer is of the usual constant current type, consisting of a three-legged laminated magnetic core, the middle leg of which is surrounded by two flat secondary coils fixed in position, and by a primary coil suspended on a rocker arm.

The d.c. reactance is tapped off the neutral of the transformer secondaries which acts the same as in the case of the storage battery charging apparatus, to store up and give out energy and hence to smooth out the waves.   The exciting transformer consists merely of a step-down transformer exciting the primary.   The secondary terminals are connected to the cathode and starting anodes.

The high tension bulb is similar to those described previously although, being of high voltage, it must necessarily have better protection for its anodes.   The entire apparatus in an assembled form for arc lighting service is shown in Fig. 3.

Fig. 3.

The large tank in the cut is filled with transformer oil and not only provides for a cooling effect on the bulb and other portions of the apparatus but also acts as a high tension insulator.   About 90° F. is the best temperature for operation, the lower limit being about 50° F.

The action or principle of operation of the mercury rectifier is a subject upon which experts seem to vary a great deal.   In fact three distinct theories are developed for its action, namely, the Steinmetz arc theory, the Thomas cathode resistance theory, and the Wagoner ionization theory.   Of these three the Wagoner theory seems to be the easiest to follow.   It maintains that the action of the rectifier tube is based on the production of negatively charged ions at the cathode spot.   These ions can be

deprived of their negative charges only at a positive electrode.
Since there exists in the tube no means of producing negative
ions at the anodes, for the very reason that there happens to
be no mercury at the anodes, the current passes in one direction
only in the arc stream. The unidirectional passage of current
is not a property of the mercury vapor (which under ordinary
circumstances is not a conductor at all), but is the result of the
employment of ion producing cathodes and anodes, which do
not produce negative ions, but merely relieve such ions of their
charges. Should an ion producing spot, or cathode be established accidentally on either anode, current would pass or arcing
would take place from one anode to the other as already described, causing a short circuit of the tube. The flow of current
in the rectifier can best be discussed with reference to Fig. 4.
Assume an instant when the terminal H of the supply transformer is positive. Then the anode A is positive and the arc is
free to flow between A & B. Following the direction of the

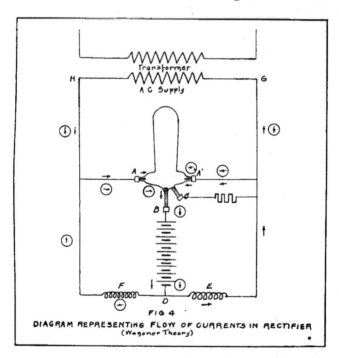

FIG 4

DIAGRAM REPRESENTING FLOW OF CURRENTS IN RECTIFIER
(Wagoner Theory)

arrows still further the current passes through the load J,
through the reactance coil E, and back to the negative terminal
on the transformer, G. A little later, when the impressed E.
M. F. falls below a sufficient value to maintain the arc against
the counter E. M. F. of the arc and load, the reactance E which
has hitherto been charging, now discharges, the discharge
current being in the same direction as formerly. This serves
to maintain the arc in the rectifier until the E. M. F. of the

supply has passed through zero, reverses, and builds up to such a value as to cause A¹ to have a sufficiently positive value to start an arc between it and the mercury cathode B.

The discharge circuit of the reactance coil E is now through the arc A¹ B instead of through the former circuit. Consequently the arc A¹ B is now supplied with current, partly from the transformer and partly from the coil E. The new circuit from the transformer is indicated by arrows enclosed in circles.

Methods of Starting.

There are several methods for starting the rectifier, or for causing the evaporation of the mercury. The common one is that of starting a spark to jump by breaking contact between the mercury in the starting anode and the cathode by shaking the tube and thus giving consequent vaporization. Or, the above is sometimes performed automatically by magnetic coils. A third in use is the aluminum starting method—where a coating of aluminum on the outside of the cathode is connected through a condenser to the anode, causing the mercury and glass at the cathode to be of opposite polarity; hence causing sparking with resultant vaporization.

Frequency and Power-Factor.

The frequencies possible on the mercury arc rectifier are really those which it is possible to apply to the transformer. However, standard outfits will operate satisfactorily on any frequency from 25 to 140 cycles, although they are usually designed for 60 cycles. The variation in frequency would, of course, depend on the other apparatus including transformers, reactance, etc. The power-factor of the system is about .9, depending on the amount of reactance in and out of the circuits.

Efficiency.

To illustrate the economical efficiency of the battery charging rectifier over the motor-generator set, let us consider an example. Suppose each to be charging a 44 cell battery, the operation being performed in accordance with conditions of time and current recommended by the manufacturer. The first part of the charge is to be at 28 amperes and 106 volts (average) for 5 hours. The efficiency of the motor-generator set at this load is 62%, and of the rectifier set at the same load, 80%.

The second part of the charge is at 12 amperes, and 108 volts (average) for two hours. In this case the efficiency of the former set is 36%, and of the rectifier 81%.

In the motor-generator set the first part of the charge requires 23.93 K. W. hours and the second part 7.2 K. W. hours from the service mains, or a total of 31.13 K. W. hours during the charge. Considering power to cost 6 cents per K. W. hour, the cost per charge amounts to $1.867.

In the case of the rectifier, the first section of the charge requires 18 K. W. hours, and the second part 3.2 K. W. hours from the service mains, making a total of 21.7 K. W. hours. At 6 cents, the cost per charge is $1.302. This shows a saving of 56½ cents per charge by using the rectifier.

Assuming the minimum life of the bulb to be 400 hours, an average of seven hours per charge allows the bulb to experience 57 charges. During this time the total saving amounts to $32.20.

Of late, rectifiers have been applied to numerous uses besides those of battery charging and constant current series arc lighting. Among them are the following: railway signals and car lighting, telegraph systems, fire and police alarm systems; motor-boats, hotel annunciators: chemical laboratory and electrolytic work, electroplating: surfacing and casting machines, elevator purposes, etc.

The many advantages of the apparatus, such as simplicity of operation, economy of initial cost, high efficiency of conversion, long life of rectifier bulbs, and fine gradations of current adjustment throughout the full range, are increased by absence of moving machinery, small space to be occupied, etc.

The outlook for the rectifier is brilliant. Its advancement has surprised even the most optimistic manufacturers. Should its construction in the future be so imposed as to allow of high commercial currents, it will undoubtedly become a central station necessity.

WATER WHEEL GOVERNORS.

E. R. FROST, B.A.Sc.

The ideal turbine governor would be one that would effect a change in output by varying only the quantity of water supplied to the wheel, thus obtaining perfect water economy by conserving unneeded water for future use. This is not possible in practice, as head water, and therefore efficiency, are usually wasted when operating a wheel under other than its normal load.

The success of the comparatively recent application of hydraulic power to the operation of alternators in parallel, and to the generation of current for electric lighting, street railway and synchronous motor loads has been largely dependent upon the possibility of obtaining close speed regulation with good water economy and without undue shock upon machinery and penstocks, while working under extremely varying loads.

While the development of automatic governing apparatus has been almost entirely experimental, remarkable results have been attained, so that some of the modern turbine governors under favorable conditions operate quite as well as modern steam engine governors.

If we consider for a moment the problem of governing a turbine we shall see that it is not at all so simple a matter as would at first appear.

The conditions of the installation have a marked effect on the difficulties to be overcome in turbine governing. If the wheel is set in an open flume having only a short draft tube, so that the water flows to the gates from every direction, the velocity of flow from any direction is very low. The quantity of water which moves at a high velocity is confined to the wheel and draft tube, and a change in the velocity and momentum due to a change in the gates produces no serious effects.

If, however, the water be brought to the wheel in a long penstock, and leaves in a long draft tube, the conditions become quite different. A large amount of energy is stored up in the moving column of water, and if an attempt is made at too rapid regulation the wheel will be left deficient in energy when more power is desired, or when the power is decreased, may produce shocks that will seriously affect the regulation or may cause serious damage to the penstock and wheel.

If the load on the wheel is decreased, the velocity of the water in the penstock must be decreased, and vice versa; work must be done on the water to accelerate it, and must be absorbed to retard it. The work required to accelerate the water must be obtained at the expense of the work done upon the wheel. Thus, when a load is thrown on the unit and the governor opens the gates, the immediate effect is a decrease in the output of the wheel, even below its original value, which is just opposite to the effect desired. Unless energy in some form is available to partially supply this deficiency, the speed of the wheel will fall considerably before readjustment to normal conditions can take place.

In the same way energy must be absorbed when the load is decreased. If this is expended on the wheel the speed will rise above normal. It may be partly dissipated by such means as relief valves or standpipes.

The water in the draft tube must also be accelerated or retarded at each change of gate at the expense of power output in exactly the same way as that in the penstock.

Since the immediate effect of the gate motion is the opposite to that intended, the governor to work properly must be made to anticipate the amount the gate must be moved to bring the speed back to normal. It must move just this far, or a little farther, and then stop and wait for the speed to return to normal.

The governor which does this is said to be "dead beat," or is compensated. Most governors of the later types, as will be shown later, are so constructed that adjustments may be made to give the degree of compensation desired. The speed of the ordinary wheel is away from normal two or three seconds, and since the best large hydraulic governors can make a full stroke in two seconds or less, it will be seen that unless the governor were compensated the gate would be entirely opened or closed long before the speed were back to normal, with the result that the wheel would race.

The different makes of governors are distinguished by their different mechanisms for compensating, and some of them have been brought to such perfection that to see one of them operating, one would almost think the machine provided with brains.

Before taking up the construction of governors it might be well to look into some of the conditions effecting regulation.

Water Hammer.—"If the closure of the gates is rapid, the column of water in the penstock is set into vibrations or oscillations, which in a long pipe under high heads is very dangerous to the wheel and penstock. If the partial gate closure is slow enough each increment in pressure is gradually dissipated along the pipe, and the hammer is avoided.

The extinction of a velocity of four feet per second at a uniform rate in one second in a pipe 1,600 feet long would create a pressure head of about 200 feet, or a total longitudinal thrust on the pipe line at each bend and upon the wheel gate of 24 inches in diameter of about 20 tons.

Hence it is seen that gate movements must be sufficiently slow to avoid dangerous oscillatory waves. This has to be decided for each plant depending on the conditions existing.

To take care of sudden pressure, relief valves and stand-pipes are often used.

In cases where the head is not very great, say up to 80 or 100 feet, stand-pipes are generally best, as they are certain of operating, and do not waste water. A stand-pipe is simply a large pipe built upright from the main penstock as close to the wheels as possible. It is open at the top, so that the water may rise and fall in it with the variation in pressure in the penstock. A stand-pipe answers two purposes: (1) As a relief valve, and (2) as a storage of energy to take care of sudden increases of load while the water is accelerating and to dissipate the excess kinetic energy in the moving column of water at a time of sudden drop in the load. For these purposes it should be as near the wheel as possible, and of ample diameter.

When the head gets over 100 feet it is impracticable to build stand-pipes owing to the cost of building a structure high enough and strong enough to withstand the wind pressure, and also because the inertia of the long column of water in the stand-pipe will give rise to strains as injurious as those they are designed to relieve.

A relief valve is arranged to open when the pressure rises above a set value and allow water to waste into the draft tube or other convenient place. An ordinary valve, such as that used for steam, would open and close so rapidly as to act like a reed upon an organ pipe, and thus maintain and increase the vibration of the mass. In order to be at all effective the valve must be dead beat. Though it must open quickly, it must close very slowly. Relief valves are now made, generally of the hydraulic type, which will open instantly and take several minutes to close, the speed of action being, of course, adjustable. Although means have thus been found for relieving the penstock of excessively high pressures, no means have yet been found for quickly accelerating the water when because of an increase in load the speed begins to fall. As already explained, when the governor opens the gates, the inertia of the mass cannot keep pace with the increased demand for water, with the result that the pressure drops, and the speed of the wheel will drop still farther. This may continue for several seconds till the water column has become sufficiently accelerated to supply the needs of the wheel.

If the governor is not made very dead beat the gates by this time will be open too wide, with the result that the speed will go too high, causing the governor to close the gates quickly, so that the relief valves will operate, which, of course, will result in very poor speed regulation.

If the governor is made dead beat enough to avoid pressure oscillation the speed regulation will be poor if the load changes are large and sudden. It is impossible for this to be otherwise, because the column of water cannot possibly alter the energy given to the wheel fast enough, whether the gates be in one position or another.

The only complete remedy for the troubles in speed regulation caused by excessive inertia of a water column is some form of by-pass valve directly connected with the water wheel gates and arranged to open when they close, and thus keep the velocity of the water nearly constant.

This would be equivalent to an impulse wheel arranged with a deflecting nozzle. It would, of course, be very wasteful of water, and in most installations could not be used.

A modification of this principle is used at Turbine, Ontario. In this case the valve, or by-pass, is indirectly controlled by the movement of the governor. The gate-controlling rack which is moved by the governor piston is extended by a piston rod having a piston on its end placed in a cylinder filled with oil. From either end of this cylinder pipes are taken to a controlling piston which operates the relief valve exhausting water from the wheel case into the draft tube. Hence any movement of the governor piston will affect the relief valve. When the governor closes quickly the valve will open, and a movement in the opposite direction will close it. As it is not desirable to have the relief valve open for small movements of the gates the cylinder is provided with an adjustable by-pass so that unless the governor makes a sudden movement of some length the oil will pass from one end of the cylinder to the other.

The small cylinder operating the valve is also provided with a by-pass so that the pressure is gradually reduced, allowing the valve to close. It was found when using this arrangement that the valve had a tendency to stay open too much, owing partly to the pressure of the water on the face of the valve. A large spring was then put behind it to insure of its closing. The by-pass, of course, regulates the speed of closing.

Governors.—To pass on now to a description of a few types of governors:

In all reaction turbines, and in all impulse turbines, with the exception of tangential wheels, the governor controls the speed by opening or closing the regulating gates, thus varying the amount of water supplied to the wheel.

As a large force is necessary to move the gates (sometimes 50,000 pounds), it is clear that they cannot be moved by fly balls alone, as in the case of a steam engine. Consequently some form of relay mechanism has to be used. That is, the movement of the balls controls some independent power for moving the gates. There are two classes of governors, viz., mechanical and hydraulic, each being classified by the type of its relay mechanism. In the mechanical type the power to move the gates is supplied by the wheel itself by means of belts, friction clutches, gears, etc. The fly balls throw into action pawls, friction gears, or other mechanical devices which will bring the relay into action. When the relay is of the hydraulic type it generally consists of a piston connected by some mechanical device to the gate rigging and moved by means of the hydraulic pressure of the water taken from the penstock or other source or by oil supplied under high pressure from a reservoir, usually the latter.

The pressure of the oil in the reservoir is kept up by pumps driven either from the wheel itself or by a separate motor. The oil used in moving the piston is exhausted into a receiver and pumped back into the supply reservoir. The relay is usually controlled by the ball governor, through the medium of a small valve, which by its motion admits the oil or water directly to the main piston or else to a secondary piston which controls a larger admission valve.

The mechanical governor is cheaper to build, and probably requires less attention than the hydraulic type, but on the other hand, it takes power from the wheel just at the time when more power is needed. It is also slower, and in the case of some wheels whose gates have a tendency to close themselves, a constant rubbing of the friction wheels is necessary to keep them open. Where the wheels operate factory machinery or machinery of that class, the mechanical governor is generally quite satisfactory and also cheaper to install.

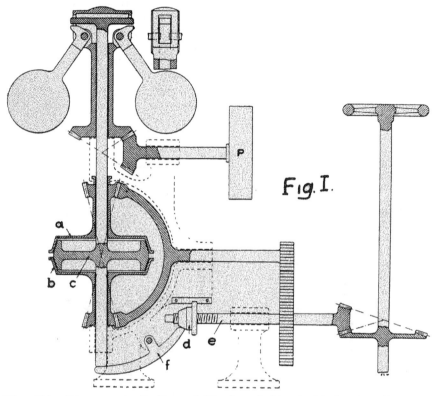

Fig. 284.—Diagramatic Section of Woodward Simple Mechanical Governor,

Fig. I shows a simple form of mechanical governor, made by the Woodward Governor Co., Rockford, Illinois.

On the upright shaft are two friction pans (a and b). These pans are loose on the shaft, the upper one being supported in position by a groove in the hub, and the lower one by an adjustable step bearing. Between these pans, and bevelled to fit them, is a double-faced friction wheel (c) which is keyed to the shaft. This shaft and friction wheel run continuously, and have a slight endwise movement. They are supported by lugs on the ball arm, and therefore rise and fall as the position of the balls varies with the speed.

When the speed is normal, the inner, or friction wheel, revolves freely between the two outer wheels or pans, which remain stationary. When a change of speed occurs the friction wheel is brought against the upper or lower pan as the speed is either slow or fast. This causes the latter to revolve, and by means of the bevel gearings, turns the gates in the proper direction until the speed is again normal. As the gate opens, the nut (d) travels along the screw (e) which is driven through gearing by the main governor shaft, and as the gate reacts, the nut (d) coming in contact with the lever (f) throws the vertical shaft upward and the governor out of commission. This governor is not dead beat, but where load changes are not very rapid or frequent, the governor works quite satisfactorily, and requires almost no attention. It may be used to advantage on exciters in small plants and for light work. The Woodward Co. also make a compensating governor of the mechanical type, but as space is limited, it will be necessary to show only one of each class.

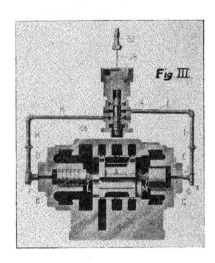

Fig. 2 shows the Lombard Governor Co.'s Type N governor.

It will be seen that the main working cylinder (1) is vertical, and that the movement of the piston up or down is transmitted through the double rack and pinions to the operating shaft. The balls control a small regulating valve, 14, Fig. 3, which admits oil from the pressure tank to the relay valve. 2. At 3 is seen the displacement cylinder with its piston connected rigidly to the rack bars.

Fig. 3 is a section of the relay valve, 2. Its method of operation is as follows:

A is the relay valve, which may be moved in either direction by the plungers B and C. Since the plunger B has one-half area of plunger C, the latter can overpower the former if the pressures in the cylinders E and D are equal. Cylinder D is connected through pipe H to the pressure supply; hence the tendency of B is always to move valve H towards the valve head G. Oil is exhausted or supplied through the pipe I. K leads to the displacement cylinder, while H is connected with the pressure tank. When valve 14 is moved by balls in one direction it admits oil from pressure tank to I, forcing A to left, which admits oil to lower side of main piston, and opens the upper side to the exhaust. The main piston rises, moving the wheel gates. In moving up, the displacement piston is carried up, creating a partial vacuum in the displacement cylinder. This draws part of the oil from I through K, and as the piston moves farther, all the oil in I is exhausted, with the result the valve, A, stops. The main piston, however, continues to move as long as valve 14 is open. When this is closed A is closed, because the liquid in cylinder, E, escapes to the displacement cylinder. When A is closed the governor comes to rest.

When 14 is moved in the opposite direction the opposite to the above takes place, so that A moves to the right till the valve, 14, takes a new position. It will be seen that any movement of the regulating valve, 14, is duplicated by the main piston, and that when valve 14 is again in its neutral position, A is immediately closed.

The anti-racing mechanism is not seen in the cut, but is arranged as follows:

The stem of valve 14 is threaded into the head of the fly balls so that any rotary motion given to it will raise or lower it. Any motion of the main piston is transmitted through a reducing motion and dash pot to a small rack engaging with a gear on the stem of valve 14.

When the speed of the wheel gets away from normal, regulating valve 14 is displaced, causing main piston to move. The reducing motion working through the dash pot and rack turns the valve stem so that the valve is closed, bringing the governor to rest to wait for the speed to return to normal. By adjusting the by-pass of the dash pot the governor may be compensated to any degree to suit the particular conditions of the installation. That is, the governor will move the gates just far enough to bring the speed back to normal, and will stop there. The Sturgess hydraulic governor uses an expansive pulley to drive the governor balls. When the main piston moves it carries with it a cam, which through a dash pot actuates a lever, expanding or contracting the pulley, thus bringing the balls back to normal speed before the wheel is actually up to normal speed.

In the case of impulse wheels, which are generally used with very high heads and long penstocks, it is impracticable to build penstocks which are strong enough to withstand the shocks due to sudden closures of gates, hence a deflecting nozzle is used. That is, the nozzle is hinged, and when it is desired to decrease the output, the nozzle is deflected from the blades. If the needle valve is set to carry peak loads there will be a waste of water at all other loads. This condition is commonly improved somewhat by adjusting the needle valve about once an hour, by means of a slow motion hand wheel, for the maximum load liable to

occur during that period. An automatic governor has lately
been invented, which adjusts the needle valve slowly automatically, so that the stream is kept on the blades. This is done by
the use of an electric motor, a connection being made which
starts the motor as soon as the nozzle deflects. The motor is
geared to move the valve slowly. In large installations employing hydraulic governors it is generally the custom now to have
one set of pumps and pressure system to supply all the governors in place of having a separate pump and tanks for each governor. This has the advantage that only one pump and pressure system have to be kept up, but on the other hand, should
this pump fail, all governors are out of commission. This is generally overcome by having a duplicate pressure stystem to take
the place of the working system while repairs are made or
while the first system is being cleaned out. The different companies making governors supply special oil, but they do not
seem to have yet found an oil which will stand up under the
continued surging around the system. A substance like cement
forms in the bottom of the tanks and piping which is very difficult to remove. Some good work might be done in experimenting along this line.

REPORT OF DUTY TRIAL ON THE SIX MILLION IMPERIAL GALLON PUMPING ENGINE AT THE HIGH LEVEL PUMPING STATION, TORONTO WATER WORKS.

ROBERT W. ANGUS, B.A.Sc.

Professor of Mechanical Engineering.

The city of Toronto has two pumping stations for the
supply of water to the city proper, exclusive of that supplied
to the residents of Toronto Island. Of these two the main
pumping station is situated at the foot of John St., close to the
bay, and all the water supplied to the city passes through
pumps in this station, the pressure being maintained at slightly
over 90 pds. per sq. in.

As the ground rises very rapidly as one proceeds north-
ward from the bay, the pressure in the northern part of the
city produced by the pumps at the main pumping station, would
be rather low and in the district near the Canadian Pacific
Railway does not much exceed 21 pds. per sq. in.

In order to maintain the proper pressure in the northern
part of the city the High Level Pumping Station was built on
Poplar Plains Road a short distance above the Canadian Pacific
Railway tracks.

The growth of the northern part of the city has been very
rapid of late years and the consequent consumption of water

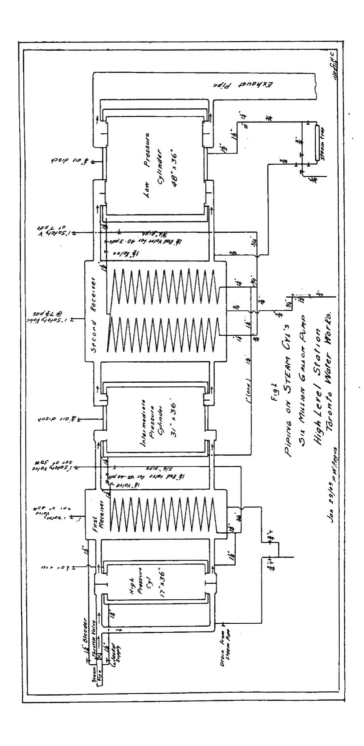

has so increased that the pumps originally installed in the station were unable to maintain a sufficiently high pressure, so that an additional pump, having a capacity of six million Imperial gallons per day against a pressure of 75 pds. per sq. in., has been installed and it is this latter pump with which this report deals.

Description of the Engine.

The engine tested is a three-cylinder, vertical, triple-expansion crank and fly-wheel pump having three single acting plungers direct-connected to the pistons of the three steam cylinders. It is designed to give a discharge of six million Imperial gallons per twenty-four hours against a discharge pressure of 75 pds. per sq. in. for domestic purposes but is also capable of giving the same discharge against a pressure of 100 pds. per sq. in. for fire purposes.

The nominal diameters of the steam cylinders as given on the working drawings, are 17 in., 31¼ in., and 48 in., respectively and the nominal diameter of all water plungers is 21¾ in., the stroke for all plungers and pistons being 36 in. For the duty trial the diameters and strokes of the water plungers were determined with great care, as is explained later, but the dimensions of the steam cylinders were not verified.

To the crosshead of the low-pressure plunger are attached the feed pump, the air pump, and an air compressor for providing compressed air in the discharge air chambers and also for the operation of the steam cylinder poppet valves if desired.

The engine has three cranks placed 120 degrees apart, the sequence being high pressure, low pressure and intermediate pressure. The crank shaft is made in two parts, which are joined together at the central crank by a sliding block which gives the shaft some flexibility without affecting its working.

Steam cylinders and piping.—A diagram of the steam piping for the engine is given at Fig. 1, which shows the main steam piping as well as that for the jackets and reheaters. Each of the cylinders is provided with a steam jacket and receivers are placed between each pair of cylinders, a reheating coil being placed in each of the receivers. The sizes and arrangement of pipes are shown on the drawing and are as follows:

(a) *Cylinder-Steam piping.*—After passing the throttle valve the steam main has two five-inch branches one of which carries steam to the top of the high pressure cylinder, the other to the bottom of the same cylinder. The exhaust from the high pressure cylinder is conveyed by two pipes, each 5 in. diameter, to the first receiver where it is reheated before being delivered to the intermediate cylinder through two pipes each 8½ in. diameter.

From the intermediate cylinder the steam passes through two pipes into the second receiver where it is again reheated before being sent to the low-pressure cylinder. After passing through the latter cylinder the steam is conveyed by the 16 in. exhaust pipe to the heater and finally to the condenser.

(b) *Jacket and reheater piping.*—The jacket supply is drawn from the main steam pipe on the boiler side of the throttle valve by means of a 1¼ in. pipe, which pipe contains a valve and connects directly with the high pressure jacket.

On leaving the high-pressure jacket the steam passes, by means of a 1¼ in. pipe, to the reheating coil in the first receiver, from which it is conveyed through a 1¼ in. pipe containing a 1¼ in. globe valve and a 1¼ in. reducing valve, (set for reducing the pressure from 150 pds. to 40 pds.), to the intermediate jacket. This 1¼ in. pipe also contains a 1 in. safety valve set at 50 pds.

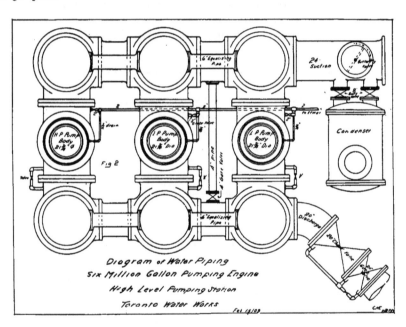

Diagram of Water Piping
Six Million Gallon Pumping Engine
High Level Pumping Station
Toronto Water Works

After passing through the intermediate jacket the steam passes through a 1 in. pipe which is enlarged to 1¼ in. into the two reheating coils in the second receiver, which coils are arranged in parallel. The steam leaves these coils through a single 1½ in. pipe, on which is a globe valve, a 1½ in. reducing valve set for reducing the pressure from 40 pds. to 3 pds., and a safety valve set at 7 pds. This 1½ in. pipe delivers the steam into the low pressure jacket.

On leaving this jacket a 1½ in. pipe, which is reduced to 1¼ in. and finally to ¾ in. delivers the steam to a trap from which it passes by a ¾ in. pipe to the sewer.

For drainage from the jackets and receivers and the reheating coil ¾ in. pipes are arranged as shown.

(c) *Other piping.*—A 1¼ in. pipe is connected from the main steam pipe on the boiler side of the throttle to the first receiver for starting up. This pipe contains a 1¼ in. globe valve.

Air discharge pipes each ¼ in. diameter and supplied with a valve are placed at the top of each cylinder jacket.

A 1 in. safety valve set at 40 pds. is placed on the first receiver and a similar valve set at 7½ pds. on the second receiver.

The high pressure cylinder has Corliss admission and exhaust valves, and on the intermediate cylinder Corliss admission valves are used, while for the exhaust for this cylinder and the admission and exhaust for the low pressure cylinder poppet valves are used.

The speed of the engine is controlled by a flyball governor which operates on the high-pressure valves only and in case the speed becomes excessive this governor also opens a valve in the condenser so as to admit air to the latter and "break" the vacuum.

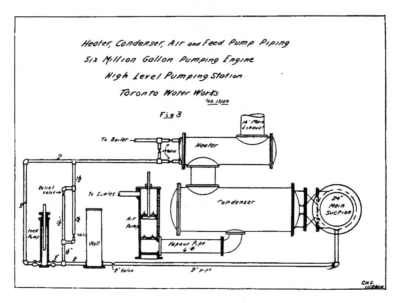

Feed Water and Condenser Piping.

The feed water and condenser piping scheme is shown on Figs. 2 and 3.

The cooling water for the condenser is taken from and again returned to the suction pipe of the engine. A butterfly valve is placed in the main 24 in. suction pipe and the cooling water is drawn from this pipe, on the side of the butterfly valve remote from the pump, through an 8 in. valve and pipe passing into the condenser. After passing through the condenser the water is returned to the suction main through an 8 in. pipe and valve but on the side of the butterfly closest to the engine. By the proper adjustment of the butterfly valve any desired proportion of the water may be sent through the condenser.

The area of the butterfly valve is about 80% of the area of the 24 in. pipe but it is never set at less than 22½° to the normal to the pipe axis and when fully open is turned parallel to the pipe axis in which case it offers practically no resistance to flow and very little water would pass through the condenser.

The exhaust steam first passes through a feed water heater and then on to the condenser. After being condensed the steam passes through a 6 in. pipe to the air pump from which it is discharged to the sewer.

The feed water is drawn from the main suction pipe through a 2 in. pipe, containing a valve, into the well. From the well a 2 in. pipe delivers the water to the feed pump from which it is discharged through a 2 in. pipe. As shown on the drawings the water may be sent through the heater or not as desired. A by-pass of 1½ in. pipe and containing valves is connected from the suction to the discharge pipe of the feed pump.

Main Water Piping.

The water enters the pumphouse through a 24 in. pipe containing a gate valve close to the wall of the room. It then passes down the south side of the engine supplying water to the suction air chambers. After passing through the pump cylinders the water is discharged at the north side of the pump through a 20 in. pipe containing a check valve and a gate valve.

A 4 in. by-pass pipe with valve may be used to connect the suction and discharge sides of the pump if desired. This pipe is placed between the intermediate and low pressure parts of the engine.

Equalizer pipes, 10 in. diameter, connect the three suction air chambers and similar pipes connect the discharge air chambers. Air is forced into the discharge air chambers by means of the air compressor attached to the low-pressure crosshead.

The remainder of the piping consists of a 2 in. pipe with valve connected to the plunger chamber and free to discharge into the sewer provided the valve is open. The valve is controlled by a wheel on the engine room main floor. There is also a 2 in. by-pass from the plunger chamber to the discharge chamber of each pump. This pipe also contains a valve which may be opened by a hand wheel on the main floor.

Pumps.

The pumps are single-acting, and corresponding to each plunger there is one suction and one discharge chamber. The valves are arranged in cages, there being for each plunger seven valve cages, each cage containing 25 valves. There are thus 525 suction valves and 525 discharge valves.

The area through each of the valves is given on the drawings as 5.95 sq. in., but this was not verified.

NOMINAL DIMENSIONS OF ENGINE AND PUMPS.

Note:—The dimensions given in the following table are all taken from the working drawings but were not verified, as they are not essential to the duty trial. The exact diameter and stroke of each plunger is given elsewhere, but the sizes given on the drawings are set down here for convenience.

1. Nominal Dimensions of the Engine.

High Pressure Cylinder:—

Diameter of piston..................... in. 17
Diameter of counterbore of cylinder.... in. 17⅛
Diameter of piston rod................ in. 4
Stroke of piston...................... in. 36
Clearance (least distance from piston to
        cylinder head)............... in. ¼

Intermediate Pressure Cylinder:—

Diameter of piston.................... in. 31¼
Diameter of counterbore of cylinder.... in. 31 5-16
Diameter of piston rod................ in. 4
Stroke of piston...................... in. 36
Clearance (least distance from piston to
        cylinder head)............... in. ¼

Low Pressure Cylinder:—

Diameter of piston.................... in. 48
Diameter of counterbore of cylinder..... in. 48⅛
Diameter of piston rod................ in. 4
Stroke of piston...................... in. 36
Clearance (least distance from piston to
        cylinder head)............... in. ¼

First Reheating Receiver:—

Diameter of two steam pipes entering.... in. 5
Diameter of two steam pipes leaving ..... in. 8½
Volume of receiver.................... cu. ft. 33.5
Size pipe in reheating coil (o.d. copper
        tubing)..................... in. 1¼
Number of coils........................ 1
Number of turns per coil............... 18
Mean diameter of coil................. in. 24
Heating surface in coil............... sq. ft. 37

Second Reheating Receiver:—

Diameter of two steam pipes entering ... in. 8½
Volume of receiver.................... cu. ft. 47

Size of pipe in reheating coils (o.d. copper
           tubing)...................... in. 1¼
Number of coils.......................... 2
Number of turns per coil.................. 22
Mean diameter of coils................. in. 28
Heating surface in two coils.......... sq. ft. 64

Condenser:—

Diameter of shell inside............... in. 36¼
Diameter of exhaust inlet.............. in. 16
Diameter of condensed steam outlet..... in. 6
Diameter of water conections.......... in. 8
Tubes No. 16 B. W. G., outside diam..... in. 1
Length of tubes between plates.......... ft. 5
Number of tubes........................ 228
Cooling surface in condenser..........sq. ft. 300

Feed water Heater:—

Diameter of shell inside............... in. 16
Length between tube plates............. in. 58
Diameter of steam inlet and outlet...... in. 16
Diameter of water connections.......... in. 2
Tubes No. 18 gauge, outside diam...... in.   ⅝
Length of tubes between plates.., ... in. 58
Number of tubes....................... 68

Air and Feed Pumps:—

Air Pump—Single Acting.

Diameter .... in. 12
Stroke .... in. 36
Diameter of inlet pipe.................. in. 6
Diameter of discharge pipe............. in. 8

Feed Pump—Single Acting.

Diameter .... in. 1¾
Stroke .... in. 36
Diameter of inlet pipe.................. in. 2
Diameter of outlet pipe................. in. 2

Air Compressor:—

Diameter .... in. 3
Stroke .... in. 36
Water jacket pipes, inlet and outlet.... in. ¾

General Dimensions:—

Length of connecting rod, centre to centre. ft. 7½
Diameter of crank shafts in fly wheel.... in. 12½
Diameter of main bearings............. in. 10½
Length of main bearings............... in. 18
Diameter of hole through shaft......... in. 3
Crank pins—diameter.................. in. 6½

Crank pins—length..................... in. 6
Crosshead pins—diameter.............. in. 6½
Crosshead pins—length................ in. 6
Diameter of steam pipe............... in.
Diameter of exhaust pipe............. in. 16
Number of flywheels...................   2
Diameter of wheels................... ft. 12 1-3
Rim 11in. and 12 in. wide, 12 in. thick.
Length of hub........................ in. 17½
Weight of each wheel................. lbs. 20000

2. Nominal Dimensions of the Pumps.

Diameter of plungers.................. in. 21¾
Stroke of plungers.................... in. 36

Valves :—

Rubber, arranged on sides and top of cages,
           secured to deck plates.

Number of valve deck plates...........   6
Number of holes for cages in each plate..   7
Total number of valve cages...........   42
Number of valves per cage.............   25
Total valves in entire pump........... 1050
Water opening in each valve........ sq. in. 5.95
(Note—one-half (525) of the above valves
   are suction, the rest discharge).

Air Chambers :—

Number of suction air chambers........   3
Number of discharge air chambers......   3

Piping :—

Suction pipe diameter................. in. 24
Discharge pipe diameter....... .... in. 20
Equalizer pipes, diameter............. in. 10
Number of distance rods to plungers.....   12
Diameter of each rod................. in. 3

The Duty Trial.

According to the specifications and contract, "The engine shall perform a duty of not less than one hundred and sixty million (160.000.000) foot pounds for each one thousand (1000) pounds of commercially dry steam used by the engine and any auxiliary pumps supplied by the contractor and operated during the duty trial. Steam containing less than 1½ per cent. of entrained water, as determined by calorimeter measurements. shall be considered as commercially dry steam. In computing the duty, the work performed by the engine shall be based upon plunger displacement. The head for computing the duty shall be that shown by an accurate pressure gauge attached to the discharge main at a point inside of the engine room and beyond the last pump, less the reading shown by a gauge attached to the supply main at or near the entrance to pump. No allowance shall be made for friction of water in pumps or pipes between the pump well and the gauge attached to the discharge main. In computing the duty, the total steam used, including that used by jackets, reheaters, and auxiliary pumps, shall be charged to the engine. The duty trial shall be of twenty-four hours' duration. The engine shall be operated continuously at the rated capacity against a head equal to 75 pds. pressure per sq. in. on the discharge main, and shall be supplied with steam of not more than 150 pds. pressure per sq. in., by gauge, at the boiler."

"The engine shall have a capacity of six million Imperial gallons in twenty-four (24) hours when operated at a plunger speed of not over 180 ft. per min., against a head equal to 75 pds. pressure per sq. in., on the pumps."

It is further stated in the general data given the contractor that the pressure in the suction main is 15 pds. per sq. in.

The trial was made as closely as possible under the contract conditions; as however, the pressure in the suction main, on account of some alterations in the city water piping system, had been raised to about 25 pds. per sq. in., the pressure on the discharge main during the test was about 85 pds. per sq. in., in order to obtain the pressure difference of 60 pds. per sq. in. between suction and discharge mains contemplated in the contract.

Weight of Steam Used.

The steam chargeable to the engine was determined by the condensation from the condenser, the jackets, the reheater, etc., and tanks were arranged, placed on scales, so that the weight of the condensed steam could be directly determined. The condensation from the condenser was measured by itself in one set of tanks, and that from the jackets and other drain pipes in a second and smaller set. The weights of condensed steam were measured every half hour, two observers checking all weights independently, setting down their results and comparing them before making the entry on the observation sheet.

Pressure.

The pressure on the discharge main was taken by an accurately calibrated Bourdon gauge, the piping leading to which was attached outside the last connecting branch from the pump. The pressure on the suction main was measured similarly by an accurately calibrated Bourdon gauge, the attachment being made close to where the suction main enters the pumping station, just outside of the condenser. The gauges themselves were placed side by side in a position where they could be conveniently read from the engine platform, about fourteen feet above the suction and discharge mains, correction being made for water column. The pressure difference was maintained as closely as possible at 60 pds. by manipulating a gate valve on the discharge main placed outside the point of attachment of the discharge gauge. During the night the pressure on the discharge main became excessive, and in order to maintain the pressure difference at the required figure it was found necessary to open a hydrant adjacent to the station.

The steam pressure at the engine was determined by an accurately calibrated Bourdon gauge, and the pressure in the calorimeter by a mercury manometer.

The pressures in the first and second receivers were taken from gauges on the gauge board. These gauges were not calibrated, but correction was made for water column. The vacuum in the exhaust pipe was taken from the vacuum gauges at the gauge board which was not calibrated.

The pressure in the steam jackets was determined from the regular gauges attached to the jacket, which were not calibrated or corrected in any way.

The barometer reading was obtained from the Observatory at intervals during the test. The result given is the average throughout the 24 hours, corrected to the height of the High Level Pumping Station, and the temperature of the engine room.

Speed.

The speed of the engine was determined by the revolution counter attached to the gauge board, the reading on this counter being checked by a second counter specially set up for the test; the counters agreed perfectly.

The Quality of the Steam.

The quality of the steam was determined by a throttling calorimeter connected to the steam main on the engine side of the throttle valve. The calorimeter worked satisfactorily throughout the test. The percentage of moisture in the steam was low, and showed very little variation throughout. The steam used by the calorimeter was not weighed.

Temperatures.

The temperatures of the engine room and of the boiler room were observed throughout the test.

The temperature of the exhaust was obtained from a thermometer placed in the exhaust pipe about eight feet below where the latter left the engine.

Indicator Diagrams.

Indicator diagrams of the steam and pump cylinders were taken at intervals during the test.

Measurements of the Plunger.

The diameters of the plungers were measured Wednesday, April 7th, the engine being shut down at about 4 p.m. for this purpose. By means of a steel tape, measurements of the circumference of each plunger at each end and at the middle were made. The plunger on the high pressure end was found to taper slightly from the bottom to the top, being largest at the bottom. In computing the diameters and areas of the plungers from this tape measurement, corrections were introduced for temperature, and for thickness of the tape, the tape itself being compared with a Government Standard.

The average diameter of each plunger determined as indicated above was used in computing the plunger displacement.

The strokes of the plungers were measured after the trial. A strip of oak 1½ inches square on which were fastened brackets and adjusting screws was firmly fixed between points on the frame and made parallel with the piston rod. A stout pointer, with a flattened end, the thickness of which was accurately measured, was then firmly clamped to one of the distance rods connecting the pump plunger to the cross head of the engine. The bracket screws were then so adjusted that the lower one just touched the under surface of the flattened pointer at the bottom of the stroke, and the upper one its upper surface at the top of the stroke. The distance between the points of the adjusting screws, less the thickness of the pointer end gave the length of the stroke. In determining the distance between the points of the screws a special bar was used whose length had been ascertained by micrometer callipers.

Calibration of Apparatus.

The steam gauges on the boiler and on the steam supply to the calorimeter, and the water gauges on the suction and discharge mains were calibrated both before and after the test by means of a Crosby gauge tester, and the results of the calibration were applied to the gauge readings in computing the results of the test; the readings were also corrected for water column, where necessary.

The jacket gauges, the first and second receiver gauges and the vacuum gauges, being those supplied by the contractor and fastened in place on the gauge board or other portion of the engine, were not calibrated, as extreme accuracy in their readings was not essential for purpose of the test; the recorded readings of these gauges have, however, been corrected for water column where necessary.

The thermometers used were not calibrated.

The scales used in measuring the weight of condensed steam were tested by the government inspector of weights and measures both before and after the trial.

The smaller scales on which the condensation from the jackets. reheaters. etc., were found to be accurate within the range of the weights measured: the larger scales on which the condensation from the condenser was weighed were found to average one pound light within the range of weights measured. A correction for this has been made.

Observations.

With the exception of the. measurement of the condensed steams. which readings, as mentioned above. were taken every half hour, and the barometer and thermometer readings, the observations were made every ten minutes.

Weather Conditions.

The weather on the afternoon of Thursday, April 8th, was sunny and warm, for the time of year, but during the night it became much colder, and there was some slight fall of snow. Friday was somewhat cold and damp.

Starting the Trial.

The trial began at 2.30 p.m. Thursday, April 8th, and ended at 2.30 p.m. Friday, April 9th, 1909. The watch used in the trial was compared with a chronometer at the beginning and end of the test, and was found to have lost approximately two seconds in the twenty-four hours.

The engine, after having been stopped for the plunger measurement on the previous evening, had been in operation for at least eight hours previous to the commencement of the trial, and was thoroughly warmed up. It ran satisfactorily throughout. with the exception of a slight vibration. caused by some of the cams operating and the valves on the low pressure cylinder not working quite smoothly.

The poppet valves on the intermediate and low pressure cylinders are arranged to be closed by springs or by air pressure: during the test the spring closure was used. the air compressor being used simply to keep air in the air chambers of the pumps.

Towards the end of the test a small leakage of steam from the high pressure cylinder jacket developed owing to a slight failure of the packing: the drip was caught. and from it the total loss due to this leak estimated. The weight was added to the total weight of water measured on the scales.

Observers.

The observers worked in eight-hour shifts. but each shift was present for about ten hours. These men were all skilled in such work, being students in the fourth year of the Faculty of Applied Science and Engineering of Toronto University.

The trial was under the direction of Professor Robert W. Angus, and was carried out by Mr. M. R. Riddell, in conjunction with Mr. W. W. Gray and Mr. J. J. Traill, all of the Faculty of Applied Science of the University of Toronto.

All calculations from the original observation sheets were made by Mr. Riddell. These calculations have been thoroughly checked.

Mr. Hill watched the test in the interest of the contractors.

The engineers and firemen at the pumping station were under the direction and control of the station engineer.

Measurements of the Pumps Used in Computations.

High-pressure Plunger—

Stroke, actual .... inches, 35.947
Diameter, actual 21.751
Displacement per revolution .... cubic feet 7.730

Intermediate-pressure Plunger

Stroke, actual .... inches 36.007
Diameter, actual ....·............. 21.757
Displacement per revolution .... cubic feet 7.747

Low-pressure Plunger—

Stroke, actual .... inches 35.972
Diameter, actual 21.760
Displacement per revolution .... cubic feet 7.742
Total displacement per revolution .... cubic feet 23.219
Total displacement per revolution ....imp. gals. 144.699
Volume of imperial gallon .... cubic inches 277.274

Observations and Results.

Date of Trial—

2.30 p.m. Thursday, April 8th, to 2.30 p.m. Friday, April 9th.
*Duration of Trial—*24 hours..

Corrected Average Pressures—

Boiler pressure by gauge, pounds per sq. inch...... 150.21
At engine pressure by gauge, pounds per sq. inch.. 148.85
In first receiver by gauge, pounds per sq. inch...... 24.04
In second receiver by gauge, pounds per sq. inch (below atmosphere) ....·........ 4.09
In intermediate jacket by gauge, pounds per sq. inch 39.08
In low pressure jacket by gauge, pounds per sq. inch 1.17
Vacuum by gauge, ins. mercury 27.39
Pressure on discharge main, pounds per sq. inch.... 85.14
Pressure on suction main, pounds per sq. inch...... 25.16
Height of centre line of discharge main above centre line of suction main at point of gauge attachment ....·...... .... feet 0.7
Corresponding pressure, pounds per sq. inch...... 0.30
Total pressure, difference on pumps, pounds per sq. inch 60.28

Barometer, average, at pump floor level and temperature, ins. mercury 29.54

Average Temperatures—

Of engine room, degrees Fahr. (lower platform)... 76
Of boiler room, degrees Fahr. 66
Of exhaust steam, deg. Fahr. 105.5

Calorimeter—

Pressure of supply steam at calorimeter, pounds per
    sq. inch 148.85
Pressure of steam in calorimeter (1 in. mercury
    =.4908 pounds per sq. inch) 1.21
Temperature in calorimeter, degrees Fahr. 298
Moisture in steam, per cent. 0.72

Speeds—

Total number of revolutions by counter 43,185
Average revolutions per minute 29.99
Average plunger speed, feet per minute.......... 179.814

Water Pumped—

Total number of revolutions 43,185
Plunger displacement per revolution, cubic feet.... 23.219
Plunger displacement per revolution, imp. gals. ... 144.699
Displacement in twenty-four hours, imp. gals 6,248,813

Work Done—

Total number of revolutions 43,185
Displacement per revolution, cubic feet 23.219
Total pressure difference on pumps, pounds per
    sq. inch 60.28
Work done per revolution, ft. pounds 201,542
Work done in twenty-four hours, ft. pounds, 8,703,603,033

Steam Used by Engine—

Total condensation from condenser, pounds....... 45,091
Total condensation from jackets, receivers, etc., lbs. 8,072
Total steam used by engine, pounds 53,163

Duty—

Work done by pump in twenty-four hours, ft.
    pounds 8,703,603,033
Steam used by engine in twenty-four hours, lbs. 53,163
Duty per thousand pounds of steam used, ft. lbs. 163,715,423
Duty required by specifications, ft lbs....... 160,000,000

POWER FACTOR CORRECTION.

L. S. ODELL, '09.

In these days A. C. power systems are becoming so common, and the attendant problems in power factor correction, so well understood, that to those familiar with the subject, any further discussion of power factor correction by the use of the synchronous motor must seem superfluous. However, the general installation of these power systems in Western Ontario has to some extent revived interest in an old subject, as is manifested by the enquiries received by manufacturers of electrical apparatus regarding the synchronous motor, its adaptability to the correction of power factor on systems already installed and the size and design of the motor required. It is in the hope that this discussion may be of some service in answering these enquiries that the same is published.

Whenever power is used from an A.C. line, whether it be through the medium of synchronous or induction motors or by means of a transformer, as is the case for lighting purposes, it is always found that the reading of the wattmeter, showing the power used, is less than the product of volts and amperes on the line. The percentage that the former is of the latter is the power factor. The energy component of the power carried by a line is the volt amperes x power factor. We have also to consider a wattless power component displaced 90 degrees in phase from the energy component, so that $P.F.^2 + W.F.^2 = 1$ where W.F. is the factor, by which we multiply the volt amperes to give the wattless power component. In a power system these wattless power components are not registered on the customer's meter, nor do they directly necessitate any increase in power at the generating station. However, a low power factor on a system means that a current much greater than that required for real power consumption is being carried, necessitating greater line capacity, larger machines, etc., and increasing the losses due to inductance, capacity losses, friction, etc. On a light load these losses may become very large, compared with the load, giving a system of very low efficiency.

Induction motors, when working, cause a lagging current. Transformers also constitute an inductive load, and create a lagging wattless component of power, but when used in connection with a lamp load, the reduction of power factor is not so serious. Now the wattless component of a synchronous motor, under certain conditions, leads the power component, so it is at once evident that synchronous and induction motors might be worked together on a system in such a way as to make the leading and lagging effects neutralize and create a power factor of practically unity.

Induction motors are the more common type in use for

general purpose work, so that usually a synchronous motor, when it is put in, is employed to correct a loss power factor due to lagging currents caused by the former type of motors. Suppose this power factor is .8. Experiment shows that as the power factor is corrected more and more nearly to unity, the capacity of the synchronous motor required to make a further change becomes relatively much greater as, for example, if a motor of 100 K.V.A. capacity were required to correct the power factor from .8 to .95, one of 200 K.V.A. capacity would probably be required to bring it up to unity. Hence it is not consistent with economy in first cost to install a motor large enough to completely correct the power factor of a system, as the last few per cent. of correction are dearly paid for in the increased K.V.A. capacity required in the motor. The better practice is to install a motor large enough to bring the power factor to .90 or .95 when the system is fully loaded, while on light loads the capacity of the motor will be sufficient to raise the P.F. still higher, and perhaps completely correct it.

The synchronous motor will do its work as a corrector of the power factor whether coupled to an external mechanical load or not. However, it is advisable to put it on such an external load if possible. Such a load will be represented in the motor by a power component in quadrature with the current used for power factor correction, giving a resultant load not much in excess of the correction load even with a considerable mechanical output. For example, in one case it was found that an external load, equal to 50 per cent. of the correction load could be carried with an increase in the required K.V.A. capacity of the motor of only 12 per cent.

The following example will serve to further illustrate the circumstances under which a synchronous motor may be installed. Suppose induction motors are now carrying a load of 800 K.W., with a power factor of .8. It is desired to determine the capacity of a motor of the former type which will correct the power factor to .95 and carry a load of 200 K.W.

Wattless current due to induction motor load at .8 power factor, $\sqrt{1-.8^2} \times {}^1/_{.8} \times 800 = 600$ K.V.A.

Total wattless current due to load of 1,000 K.W. at .95 power factor $= \sqrt{1-.95^2} \times {}^1/_{.95} \times 1,000 = 336$ K.V.A.

Hence the synchronous motor must furnish the difference between these two results, the former being the initial lagging component and the latter the lagging component after correction. Combining this difference of 264 K.V.A. with the mechanical load of 200 K.V.A. in quadrature we have:

$\sqrt{264^2 + 200^2} = 332$ K.V.A. = total capacity of the synchronous motor.

Then the power factor at which the synchronous motor will operate as $\dfrac{200}{332} = .6$ approximately.

Hence in above case a synchronous motor of 332 K.V.A. capacity must be installed, and the excitation so controlled that it will operate under a power factor of .6.

The reason for this adjustment of excitation is as follows: Such a motor when operating under any particular impressed voltage, sustains a constant magnetic flux, no matter what the exciting current may be. A lagging or leading induced armature current is set up, this induced current being such as, when combined with the field excitation, will give a constant magnetizing force. When the fields are over excited a leading current will flow in the armature, and hence a synchronous motor must always be over excited when used to correct a power factor which is held down by lagging currents. When operating a synchronous motor, as explained above an increase in the field excitation will increase the leading current it will draw from the line, while decreasing the field excitation will decrease this leading current until when this leading current falls to zero the motor is operating under unity power factor. A further decrease in excitation would cause the machine to draw a lagging current from the line, so that if necessary a synchronous motor could be used to correct power factors for either leading or lagging currents.

Certain important characteristics effect the suitability of a motor for this class of work. Any AC generator would operate as a synchronous motor, but used for correction of power factors held down by lagging currents, might not be satisfactory. These machines are over excited, and the excitation increases directly as the current, while the heating effects in the field coils increase as the square of the current. Hence machines of low frequency where the number of poles is small and which often give trouble by heating when used as generators, will not do at all, owing to the high temperature in the fields. A machine with a small air gap decreases the amount of excitation necessary. Also the weaker the armature, or in other words, the fewer the coils in it, and the lower the resistance of those coils, the greater will be the induced leading current for the same increase in field excitation. Hence the characteristics of a good machine for power factor correction are a large number of low resistance fields, a small air gap and a weak armature. Such a machine, however, is not likely to be satisfactory as a generator.

Throughout the country there are a great many induction motors operating on power factors of from .5 to .7, and in many cases power consumers are being penalized by the companies, who insist on a power factor of .9 to secure the lowest rates. The synchronous motor offers a means of correction of this difficulty. In many cases, however, the saving will not justify the expense entailed, especially where there is the employment of skilled labor to be reckoned on along with the installation. In other cases, where the load is taken off at only a short distance from the generator, the initial expense is again prohibitive, as the increased generator capacity is the only saving effected. However, on long transmission lines where a low power factor necessitates increased capacity of generators, step up and step down transformers and all other apparatus in the system there is no question as to the advisability of using some means to balance up the lagging and leading currents and hold the power factor as near unity as possible.

THE ENGINEERING SOCIETY

The November sectional meetings of the society were held on Wednesday, the 16th. J. W. Nelson, B.A. Sc., addressed the civil section on "Surveying the Alaska Boundary." His paper was well illustrated by a series of 150 slides, and being well acquainted with his subject, he said a great deal to relieve the impression that the country was an uninteresting one. "Water Wheel Governors," by E. R. Frost, B.A. Sc., was the subject at the mechanical and electrical meeting. The paper appears in full elsewhere in this number of "Applied Science." The miners and chemists were favored with an address on "Common Food Stuffs; Their Manufacture and Adulteration," by L. J. Rogers, B.A. Sc. All the meetings were well attended.

On Wednesday, Nov. 9th, the men of the third year journeyed to the Lackawanna Steel Plant, Buffalo, under the direction of Mr. T. R. Loudon and Mr. J. A. Stiles. In parties of ten and twelve, they were guided through the works, and given an excellent chance of seeing the various processes, from the unloading of the ore at the docks, to the loading of the finished product for shipment. The entire series of processes covers a period of about fifteen hours, and requires the services of some 7,000 men.

Another beneficial excursion, this one under the supervision of Mr. Murphy, vice-president of the civil section, was that to Centre Island, where the new filtration plant is in the course of construction. Mr. Longley, the engineer in charge, spared no pains in guiding and instructing the men as to what is being accomplished there. He not only went into the details of the plant, but also explained many of the outside aqueous problems.

NOTICE.

The graduates in and around Toronto should keep in mind a special meeting of the Engineering Society on Monday evening, Dec. 12th. Mr. Isham Randolph, of Chicago, is to give an illustrated address on the "Ship and Sanitary Canal of Chicago." To hear Mr. Randolph, one of the foremost consulting engineers in America, is an opportunity open to all who can possibly attend.

APPLIED SCIENCE

INCORPORATED WITH

Transactions of the University of Toronto Engineering Society

DEVOTED TO THE INTERESTS OF ENGINEERING, ARCHITECTURE
AND APPLIED CHEMISTRY AT THE UNIVERSITY OF TORONTO.

Published monthly during the College year by the University of Toronto Engineering Society

EDITORIAL

Enclosed with this number, every reader of "Applied
Science" will receive a post card for the attention of the graduates asking for address and particulars regarding employment.
This card should be filled in, returned, and placed upon the
Secretary's desk for reference. In thus accomplishing his part,
the graduate may expect the requested information to be always
within easy reach when occasion calls for it. It matters not
whether the enquiry be from a fellow graduate renewing ac-
quaintances with old classmates, or from a firm in search of
professional assistance, the result will be of advantage to the
graduate whose whereabouts is on record at the most convenient
centre. There are, on file, records of engineering experience of
a large number of men, but, in numerous cases, these are without present addresses, and present employment, and are consequently, not up-to-date. The employment bureau has been
successful in placing almost every man who applied to it for

employment, and has often been in receipt of more enquiries for men than of applications for work. This points to the fact that the bureau is receiving recognition by engineering firms, as a fit and proper place to apply for technical men, and the younger graduate should not fail to obtain from it the assistance it is intended to convey. His best move is to assure himself that the secretary of the faculty has his name, address and professional experience to date. In short, the closer relations the graduate has with his college the better, from every point of view, no matter whether satisfactorily employed or not.

LEONARD.*

To the Editor of Applied Science:—

Sir:—What is an engineer? Leonard says he is a man who does things.† The man on the street and the newspapers say that an engineer is one who drives engines; more particularly locomotive engines, for the driver of mill engines is a "stationary engineer." The unqualified term belongs to the driver of the locomotive, that quiet, unassuming, but all-important factor in modern transportation, the big man of the train equipment that the public never tips nor sees. Other men are mechanical engineers, civil engineers, mining engineers, and so on, but to the public the mechanical engineer is a man in overalls, greasy-handed and black-marked as to the nose and eyes, who works around machines, the civil engineer is a man with a queer brass instrument on three legs, and the mining engineer is a bad type of crook associated with wild mining flotations. Is it possible that in these days there can be thinking people who have such ideas? It is more than possible, it is an everyday experience. Ask the lawyers who, as a general thing, are our best-posted citizens, and unless they have been on a case where technical witnesses of experience have enlightened them, they will show only hazy, unformed ideas as to the real status or aims of technically trained Engineers. They do not distinguish between them and the mechanic or the plumber. They seldom think of calling them in to be guided by them; they employ them to do certain things for them, to measure or construct, much as they would have a druggist concoct a prescription, and they pay him accordingly. And the engineer does the work and is paid accordingly and says nothing.

The Engineer—the technically trained man—does things, does everything that makes for health and growth and prosperity and lets it go at that. And the world at large sees no more of him and thinks no more about him than does the Pullman passenger and the driver that has brought his train through all the hazards of the trip.

*R. W. Leonard was appointed to the Board of Governors of the University of Toronto last Spring.

†"The Scope of Engineering in Canada," by R. W. Leonard, "Applied Science," Nov., 1910.

The Engineer is so busy with his work, so interested in his things that he neglects his fellowman. The Churchman, the Doctor, the Lawyer, our professional men, are wiser, they pay more attention to their fellowman than they do to their work, or perhaps it would be truer to say that their eye is first on their fellow. They advise, they guide, they control (perhaps they cajole, flatter or bulldoze), and their fellowman, being but a short time removed from childhood or tribal custom, places them on a pedestal and is glad to be a parishioner, or a patient, or a client. We care nothing for such matters; we want our work, our pride it is to do our work, and all we ask of our fellow is to stand aside and let us do this work. And so it is that we who do the nation's real work amount to but little in the public estimate. We are neither known nor understood. There are in Canada a few, a very few, exceptions that go to prove the rule. Sir Sandford Fleming is an example. How many of the younger generation of engineers take any active interest in their fellows? Very, very few. How many take part in public affairs off the beaten path of their work? There is not one of prominence, unless it be Leonard. He is a peculiarly all-round man. He began with the military training of the R.M.C. In his early work he made a marked success in railroad construction, and later in water power development. He was always interested and generally dabbling in mining. Cobalt was his opportunity and he took it. His Coniagas success was much more the work of the experienced calculating financier than the luck of the explorer. He has always been keen in military affairs. The Leonard Gig is an accepted part of field equipment. And now he is an assured financial success. He has money. That must be writ large, for despite all we may say or pretend to think to the contrary, money counts in these days. And be it noted he made his money as an engineer, cleanly. And still he remains an engineer, taking a keener interest in his engineering .work than in anything else. And on top of all this he is stepping out to take a part in his fellows' affairs; to take an active, disinterested part in the larger affairs of the nation. As a graduate of the school I am particularly glad that his first energies are with Toronto University. The School needs him. We all recognize the advantages of a special pleader, and the School is no doubt in need of a counter-balance to the urgent calls upon the funds from Arts and Medicine and the apparently innumerable new faculties. But personally I should be sorry to see Leonard either become or be looked upon as a special pleader for the material growth of the School. His is a larger function. It is a safe bet that the Engineer and his aims are as little understood in the University councils outside of the Faculty of Engineering as they are by the public. It is up to Leonard to tackle the larger education, the education of those in control. Yours etc..

AN OLD GRADUATE.

HARRY MILL LANCASTER.

Since the last issue of "Applied Science" we learn with great regret of the resignation from the staff of Applied Chemistry of Mr. H. M. Lancaster. Mr. Lancaster's academic career has been marked with singular ability; coming from Woodstock Collegiate Institute with a first Edward Blake scholarship in Science and a second in Science and Mathematics, he entered the Arts Honor courses of Physics and Chemistry and Chemistry and Mineralogy, with the class of '05. In both these courses he carried off the scholarships in his freshman year. Full first-class honor standing in his first three years enabled him to transfer to the fourth year in the Department of Applied Chemistry, where he selected the Sanitary and Forensic option. Since his graduation in 1906 with the degree of B.A. Sc., with honors, Mr. Lancaster has been connected continuously with the Department of Applied Chemistry, first as fellow, and from 1907 till the time of his resignation, as demonstrator. In addition to his academic work, he pursued many outside interests, always, however, directing his special attention to toxicological investigations and food and water analysis. This particular line of work, in which he had the good fortune of personal association with Dr. Ellis, fitted him ideally to assume the duties his present position entail. His co-workers on the staff and his students unite in wishing him every success in the responsible post to which he has been appointed.

BOOK REVIEWS

STANDARD HANDBOOK FOR ELECTRICAL ENGINEERS.

(McGraw-Hill Book Co., Third Edition; Leather).

One of the outstanding needs of every student of electrical matters is a handbook which embodies in convenient form a large collection of useful data accompanied by sufficient explanation and general information to enable him to properly use the data. Such a companion is the "Standard Handbook."

The men who have compiled the various sections of the book are engineers of standing, who have made critical study of the subject in hand. It is natural, therefore, to find that necessary explanations involving complexity and formulae are not avoided in the Standard as in some other handbooks. This feature is of value to students whose point of view demands understanding of means and methods rather than accumulation of data and results alone.

Sections 1, 2, 3 include in concentrated form information on units, methods and calculations which appeals to students because it bears directly on one large phase of their studies. Section 4, on properties of materials, supplements the previous sections by providing considerable definite data for use therewith.

Sections 5 to 9 refer to the theory, design, construction, operating properties and testing of generators, motors, transformers, batteries, magnets, rheostats, etc. There is presented a surprisingly large quantity and variety of information. The treatment is very complete, considering the space available.

Sections 10 to 18 deal with the problems of electricity as applied to central station plants of all kinds, transmission and distribution, under various conditions, electric lighting in general, electric traction in its many phases, from 600 volt city services to 11,000 volt heavy traction problems, electro-chemical problems, with their industrial applications and practice, telephony, telegraphy, etc., etc. Each section is an excellent text and handbook on theory and practice in its field.

Section 19 is useful, as it covers the standardization rules and recommendations of the best engineering associations in America. Section 20 includes a number of mathematical tables and statistical data relative to electrical industries.

When in search of information the reader is assisted in locating matter in point by the use of heavy type in the text for words in each paragraph, which suggest the subject matter thereof. The index seems complete, but would be much more effective if cross-indexing were considerably increased.

The "Standard Handbook" will be a permanently valuable addition to the library of any student. *Reviewed by H. W. Price, B.A. Sc., Dept. of Electrical Engineering.*

WHAT OUR GRADUATES ARE DOING.

W. H. Munro, '04, has been appointed manager of the allied companies: the Peterboro' Light and Power Co., the Peterboro' Radial Railway, and the Auburn Power Co., of Peterboro', after having been engaged for the past six years in hydro-electric work.

H. F. H. Hertzberg, '07, is with the Trussed Concrete Steel Co., of Canada, in their Winnipeg office.

W. M. Bristol, '05, is on the staff of the Canadian Westinghouse Co., in their Halifax branch.

Fred H. Moody, '08, has been appointed associate editor of "Machinery," published by the Industrial Press, New York.

P. T. Kirwan, '10, and A. V. Delaporte, '10, have been appointed Fellows in Chemistry for the session 1910-11.

L. J. Rogers, '07, who has for some time been chemist for the Pure Gold Manufacturing Co., succeeds Mr. Lancaster as demonstrator in Chemistry.

Willis Maclachlan, '06, previously construction engineer in Hydro-electric work, Niagara, has been appointed city engineer for London, Ontario.

G. R. Workman, '10, lately of the Canadian Bridge Co., has accepted a position with the Laurentide Paper Co., Limited, of Grand Mère, Quebec.

Index to Advertisers

Applied Science

INCORPORATED WITH

TRANSACTIONS OF THE UNIVERSITY OF TORONTO ENGINEERING SOCIETY

Old Series Vol. 23   JANUARY, 1911   New Series Vol. IV. No. 3

STRESSES IN CIRCULAR RINGS WITH INTERNAL WATER PRESSURE.

T. H. Hogg, B.A. Sc.

The subject of this paper is the development of the stresses set up in a circular ring under internal water pressure.

The development of the theory is along similar lines to the work of C. W. Filkins and E. J. Fort, who developed the stresses in circular rings due to the weight of the rings themselves, and whose work is published in the "Transactions of the Association of Civil Engineers of Cornell University" for 1896. A reference to similar work is also made in a paper by Mr. Muller in the "Engineering Record" for May 1st, 1909.

The results contained in this paper were arrived at while the writer was working under the direction of Mr. R. D. Johnson of Niagara Falls. It is suggested that Mr. Johnson's article on the Hydrostatic Cord, which appeared in the April 1st, 1910, issue of the Canadian Engineer be read in connection with this paper; as the work contained herein is a development of some of the formulae used by him.

It is usually assumed in figuring the stresses due to internal water pressure in circular rings or pipes, lying on their sides, that it is quite sufficient to take only the tension induced in the shell into account, and that the bending moment may be neglected. This is only true where the pressure head is great compared to the diameter of the pipe. It is easily seen that where this condition does not exist there is a much greater pressure at the bottom of the pipe than at the top, and this may cause relatively great bending moments in the shell.

In the following discussion, the ring is assumed supported on a knife edge, and water pressure is assumed level with the crown of the pipe. The analysis also assumes a thin ring of homogeneous material having a constant modulus of elasticity, and that the changes from a circular form will have little effect upon the dimensions of the ring.

The following nomenclature will be used throughout the discussion:—

H=head of water above top of pipe.

r=radius of circular pipe.
M=bending moment in pipe.
M =bending moment in pipe at top.
M_D=bending moment in pipe at bottom.
M_{max}=bending moment in pipe at maximum point on side.
T=tension in pipe.
T_B=tension in pipe at top.
T_D=tension in pipe at bottom.
ϕ=angle to different points (expressed in radians).
ϕ_{max}=angle to maximum bending moment point on side.
ϕ_u=angle to upper node point (point of no bending moment).
ϕ_1=angle to lower node point (point of no bending moment).
γ=weight of cu. ft. of water.
J=shear at any point.
J_B=shear at top.
J_D=shear at bottom.
$d\delta$=arm of bending moment.

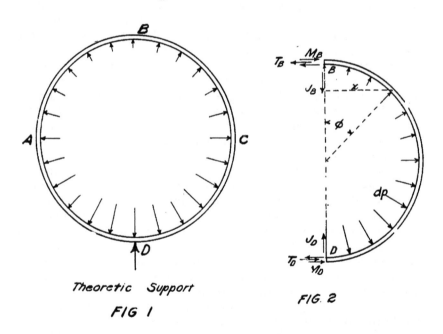

Theoretic Support

FIG 1

FIG. 2

We will assume that the reader is familiar with the three general formulae for arch-ribs.

$$\int_D^B \frac{Mds}{EI} - \int_D^B d\phi \; ; \; \int_D^B \frac{Myds}{EI} - \int_D^B dx \; ;$$

$$\int_B^B \frac{Mxds}{EI} \qquad \int_D^B dy .$$

The ring is a continuous curved beam to which these equations will apply.

The forces acting upon the pipe may be appreciated by looking at Fig. 1. If we cut the pipe at B and D and consider the forces acting on the section to the right, we obtain the system of forces shown in Fig. 2.

Taking the centre of moments at the neutral axis at D we obtain:

$$M_D = M_B - 2\,T_B\,r + \int_0^\pi (1 - \cos \phi)\, d\,p \sin \phi$$

now

$$d\,p = \gamma\,r^2\,(1 - \cos \phi)\, d\,\phi\,; \text{ and } \sin \phi\, d\,p = \gamma r^2\,(1 - \cos) \sin \phi\, d\,\phi$$

Therefore

$$M_D = M_B - 2\,T_B\,r + \int_0^\pi \gamma\,r^3\,(1 - \cos \phi)\sin \phi\, d\,\phi$$

Integrating we obtain:

$$M_D = M_B - 2\,T_B\,r + 2\,\gamma\,r^3$$

Now consider as a free body that portion of pipe shown in Fig. 3.

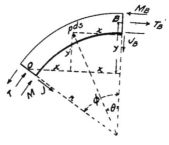

FIG 3

$$E\,I\,\Delta\,y = \int M\,x\,d\,s\,;\; ds = r\,d\,\phi = r\,d\,\theta$$

$$-\,M = M_B - T_B\,y + \int_0^B p\,ds\,x$$

$$p\,ds = \gamma\,r^2\,(1 - \cos \theta)\, d\,\theta$$

$$x = r\,\sin(\phi - \theta)$$

$$y = r\,(1 - \cos \phi).$$

Therefore

$$-\,M = M_B - T_B\,r\,(1 - \cos \phi) + \int_0^\phi \gamma\,r^3\,(1 \quad \cos \theta)\sin(\phi - \theta)\, d\,\theta$$

$$= M_B - T_B\,r + T_B\,r\cos \phi + \gamma\,r^3\int_0^\phi (1 - \cos \theta)\sin(\phi - \theta)\, d\,\theta$$

$$= M_B - T_B\,r + T_B\,r\cos \phi + \gamma\,r^3\,(1 - \cos \phi - \tfrac{1}{2}\,\phi \sin \phi)$$

Therefore

$$E I \Delta y = \int M x \, ds = \int M r^2 \sin \phi \, d\phi$$

$$= - \int_0^\pi [(M_B r^2 - T_B r^3 + \gamma r^5) \sin \phi \, d\phi + (T_B r^3 - \gamma r^5) \sin \phi$$
$$\cos \phi \, d\phi - \tfrac{1}{2} \gamma r^5 \phi \sin^2 \phi \, d\phi$$

Solving above equation we obtain:

$$E I \Delta y = - 2 (M_B r^2 - T_B r^3 + \gamma r^5) + \frac{\gamma r^5 \pi^2}{8}$$

Also $E I \Delta x = \int_D^B M (ds) y = 0$ since B has not moved hori-

zontally and since its tangent is horizontal.

Now $y = r (1 - \cos \phi)$; $d s = r d \phi$.

Therefore $E I \Delta x = r \int_0^{\pi r} M d s - r^2 \int_0^\pi M \cos \phi \, d\phi$

$$0 = r \int_0^{\pi r} M d s - r^2 \int_0^\pi M \cos \phi \, d\phi.$$

Substitute value of M found above and solve.

Therefore $- \int_0^\pi M \cos \phi \, d\phi = \int_0^\pi [(M_B - T_B r + \gamma r^3) \cos \phi \, d\phi$
$$+ (T_B r - \gamma r^3) \cos^2 \phi \, d\phi - \tfrac{1}{2} \gamma r^3 \phi \sin \phi \cos \phi \, d\phi]$$

Therefore $- \int_0^\pi M \cos \phi \, d\phi = \frac{\pi}{2} \left\{ T_B r - \gamma r^3 \right\} - \frac{\gamma r^3 \pi}{8}$

or $\int_D^B M y \, ds = \frac{\pi r^3}{2} \left\{ T_B - \frac{3 \gamma r^2}{8} \right\} = 0$

Therefore $T_B = \frac{3}{4} \gamma r^2$

Now $\int_C^B M d s = 0 = \int_0^\pi [M_B r d \phi - T_B r^2(d \phi - \cos \phi$
$$d \phi) + \gamma r^4(d \phi - \cos \phi \, d \phi - \frac{1}{2} \phi \sin \phi \, d \phi)]$$

whence, solving above,

$$M_B = T_B r - \frac{1}{2} \gamma r^3$$

Therefore $M_B = \frac{\gamma r^3}{4}$

Now $M_D = M_B + 2 \gamma r^3 - 2 T_B r$

$$\frac{\gamma r^3}{4} - 2 \gamma r^3 - \frac{3 \gamma r^3}{2}$$

$$= - \frac{3}{4} \gamma r^3$$

Now $- M = M_B - T_B r + T_B r \cos \phi + \gamma r^3 (1 - \cos \phi -$
$\dfrac{1}{2} \phi \sin \phi) = \gamma r^3 \left(\dfrac{1}{2} - \dfrac{1}{4} \cos \phi - \dfrac{1}{2} \phi \sin \phi \right)$

which is the general expression for the bending moment.

To find the shortening of the vertical diameter substitute the values of M_B and T_B in the equation.

$$E I \Delta y = - 2 (M_B r^2 - T_B r^3 + \gamma r^5) - \dfrac{\gamma r^5 \pi^2}{8}$$

We obtain: $E I \Delta y = - \gamma r^5 + \dfrac{\gamma r^5 \pi^2}{8}$

$$\Delta y = \dfrac{\gamma r^5}{E I} \left(\dfrac{\pi^2}{8} - 1 \right) = + .2337 \dfrac{\gamma r^5}{E I}$$

To find the lengthening of horizonthal diameter

$$E I \Delta x = \int^B M \, (ds) \, y$$

Therefore

$- E I \Delta x = \int_0^{\frac{\pi}{2}} [(M_B - T_B r - \gamma r^3) r^2 (1 - \cos \phi) \, d\phi$

$+ (T_B r - \gamma r^3) r^2 (\cos \phi - \cos^2 \phi) \, d\phi - \dfrac{1}{2} \gamma r^3 (1 -$

$\cos \phi) \phi \sin \phi \, d\phi$

Solving $- E I \Delta x = \dfrac{\gamma r^5}{4} (1.5 \pi - 5)$

$= - .0719 \gamma r^5$

Therefore $\Delta x = .0719 \dfrac{\gamma r^5}{E I}$

The whole change in horizontal diameter

$= 2 \Delta x = .1438 \dfrac{\gamma r^5}{E I}$

In order to find the value of ϕ where the moment is a maximum, place $\dfrac{d M}{d \phi} = 0$

Therefore $T_B r d \cos \phi - \gamma r^3 d \cos \phi - \dfrac{1}{2} \gamma r^3 d \phi \sin \phi = 0$

Differentiating and simplifying we obtain:

$$\phi \cot \phi = \dfrac{1}{2}$$

$- 2 \phi = \tan \phi$

By trial we find $\phi = 105° - 13' - 45''.4$; also $\phi = 0$.

Now, when $\phi = 0$ the maximum moment $M_R = \dfrac{\gamma r^3}{4}$

When $\phi=105°-13'-45".4$ substitute in exrpession for moment

$$- M = M_{\text{B}} - T\ r + T_{\text{R}}\ r \cos \phi + \gamma\ r^3 \left(1 - \cos \phi - \frac{1}{2}\ \phi \sin \phi\right)$$

and we obtain $M = - .3203\ \gamma\ r^3$

Assembling and comparing the values of the moments found we see that $M_{\text{B}} = \div \dfrac{\gamma\ r^3}{4}$; $M_{\text{D}} = \top \dfrac{3\ \gamma\ r^3}{4}$

and $M_{\text{Max}} = - .3203\ \gamma\ r^3$

Therefore, we see that the moment of the stress couple changes sign (i.e., passes through zero) between M_{B} and M_{Max} and between M_{Max} and M_{D}

Place M in general equation equal to zero and solve for ϕ

$$\gamma\ r^3 \left(\frac{1}{2} - \frac{1}{4} \cos \phi - \frac{1}{2}\ \phi \sin \phi\right) = M = 0$$

Therefore, $\phi \sin \phi + \dfrac{1}{2} \cos \phi = 1$

or $\phi=50°-36'-45"$,
also $\phi=146°-19'-25"$.

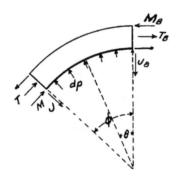

FIG. 4

Now, considering Fig. 4, we see that value of
$$d\ p = \gamma\ r^2 \left(1 - \cos \theta\right) d\ \theta.$$

Therefore the sum of the vertical components of all the $d\ p$'s between

$$O \text{ and } \phi = \int_0^\phi d\ p \cos \theta = \int_0^\phi \gamma\ r^2 \left(1 \cos \theta\right) \cos \theta\ d\ \theta$$

$$= \gamma\ r^2 \left(\sin \phi - \frac{1}{2}\ \phi - \frac{1}{4} \sin 2\ \phi\right)$$

And the sum of the horizontol components of all the $d\ p$'s between O and ϕ

$$= \int_0^\phi d p \sin \theta = \int_0^\phi \gamma \, r^2 \, (1 - \cos \theta) \sin \theta \, d \theta$$

$$= - \gamma \, r^2 \left(\cos \theta \, \phi - 1 + \frac{\sin^2 \phi}{2} \right)$$

In Fig. 4 place vertical components of the acting forces equal to zero.

Therefore,

$$J \cos \phi + T \sin \phi = \gamma \, r^2 \left(\sin \phi - \frac{1}{2} \phi - \frac{1}{2} \sin \phi \cos \phi \right)$$

Placing horizontal components equal to zero,

$$J \sin \phi - T \cos \phi = \frac{3}{4} \gamma \, r^2 - \gamma \, r^2 \cos \phi + \gamma \, r^2 - \frac{1}{2} \gamma \, r^2 \sin {}^2\phi$$

solving these equations for T and J we obtain:

$$T = \gamma \, r^2 \left(1 - \frac{1}{2} \phi \sin \phi - \phi - \frac{1}{4} \cos \phi \right)$$

$$J = \gamma \, r^2 \left(\frac{1}{2} \phi \cos \phi + \frac{1}{4} \sin \phi \right)$$

The above discussion has assumed that the pipe is just filled with water to the top. We can easily see that the bending moments are not affected by the water pressure after the top is passed. In other words, the bending moments induced in the shell are caused by the difference in pressure between the top and any other point chosen; and this difference remains constant for any head above the top. The tension in the shell, however, varies directly with the head and is equal to $H \, \gamma \, r$, where H is the head on the top of the pipe.

This amount must be added to the above obtained value of the tension to obtain the general expression.

Collecting the formulae we have

$$M = \gamma \, r^2 \left(\frac{1}{2} - \frac{1}{4} \cos \phi \, \frac{1}{2} \phi \sin \phi \right)$$

$$T = H \, \gamma \, r + \gamma \, r^2 \left(1 - \frac{1}{4} \cos \phi - \frac{1}{2} \phi \sin \phi \right)$$

$$J = \gamma \, r^2 \left(\frac{1}{2} \phi \cos \phi + \frac{1}{4} \sin \phi \right)$$

$$M_B = \frac{1}{4} \gamma \, r^3$$

$$M_D = \frac{3}{4} \gamma \, r^3$$

$$M_{Max} = - .3203 \, \gamma \, r^3$$

$$T_B = H \, \gamma \, r + \frac{3}{4} \gamma \, r^2$$

$$T_D = H \, \gamma \, r + \frac{5}{4} \gamma \, r^2$$

$$\phi_{Max} = 105° — 13' — 45''.4$$

$$\phi_u = 50° — 36' — 45''$$

$$\phi_L = 146° — 19' — 25''$$

$$\Delta x = .0719 \ \frac{\gamma \, r^5}{E \, I}$$

$$\Delta y = .2337 \ \frac{\gamma \, r^5}{E \, I}$$

In order to show the results clearly let us plot a bending moment diagram. We can always express the bending moment in terms of the tension multiplied by the arm of the moment. Therefore, if we divide the bending moment at any point by the tension at the same point, we obtain the arm of the bending moment at that point.

Let $d \, \rho = \dfrac{M}{T}$ = arm of bending moment.

$$= \frac{\gamma \, r^3 \left(\dfrac{1}{2} — \dfrac{1}{4} \cos \phi \dfrac{1}{2} \, \phi \sin \phi \right)}{\gamma \, r \left\{ H + r \left(1 — \dfrac{1}{4} \cos \phi — \dfrac{1}{2} \, \phi \sin \phi \right) \right\}}$$

$$= \frac{r \left(\dfrac{1}{2} + k \right)}{\dfrac{H + r}{r} + k} \quad \text{where } k = — \dfrac{1}{4} \cos \phi — \dfrac{1}{2} \, \phi \sin \phi$$

Plotting the values of $d \, \delta$ radially, for different values of ϕ we can represent the bending moment as shown in Fig. 5. This curve is the equilibrium polygon for the assumed conditions.

The above formulae taken together with the formulae for the weight of the shell itself, a reference to which was made at the beginning of this paper, are extremely useful; in fact, absolutely necessary to the correct design of concrete pressure pipes of large diameter.

On account of the node points (points of zero bending moment) and points of maximum bending moment occurring at the same points, both for the weight of the water itself and the weight of the shell, and because the bending moments due to both causes are exactly proportional at all points on the circumference it follows that they may be readily combined. With the above formulae, therefore, and those for the weight of the shell the design of pressure pipes of large diameter becomes a comparatively simple matter.

The above method of analysis; based on the general formulae for continuous arch girders, may also be used to determine the stresses induced in the shell for different loads applied on the top and sides, such as back-fill, etc.

The writer is sorry that lack of time prevents him from entering more fully into the methods of combining the several

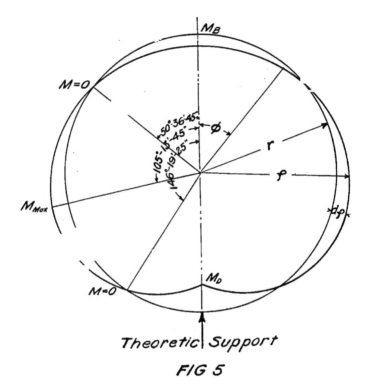

FIG 5

stresses due to weight of shell, weight of water, water pressure, back-fill and top loads, as many interesting problems arise in the application to an actual design.

RADIO-TELEGRAPHY.

L. T. RUTLEDGE, B.A.Sc.

All signalling at a distance necessitates the use of three distinct parts: the device which produces the signal, that which carries it, and that which receives it. These three essentials we may call the sender or transmitter, the line and the receiver. Now, in the case of wireless telegraphy, something in space is the medium of propagation, and it is the object of the writer to discuss the conditions which make it possible for telegrahpic messages to be sent from one continent to the other with seemingly nothing passing from one to the other.

The first point in this connection is to get a clear idea of the fields that are set up in space by electric currents and charges. It is well known that a current, even in a straight conductor, sets up an invisible magnetic field in the space surrounding it. In this case the field consists of magnetic lines that surround the conductor. Now let us briefly consider the state of the space that lies between any surface that is charged with a positive charge and the neighboring surface that is charged negatively. The dielectric, whether air or glass, between the two surfaces that are charged, is in a state of electric strain, there being electric lines of force through it from positive to negative, and these constitute an electric field.

Now consider a circuit in which there is an oscillating current. Fig. 1 represents a simple condenser made of two flat metal plates, with a layer of air between them. They are joined by a simple circuit made of bent brass rods, ending in polished balls with a spark gap between them, so as to be able to start oscillations by connecting to a suitable source, such as an induction coil. Just before a spark passes, one of the metal plates is highly positive and the other highly negative, and then there will be an electric field across the air space between them. When the spark occurs, the positive charge rushes round the circuit into the plate that was negative and then surges back again with incredible rapidity, oscillating along the conducting circuit. Each such rush constitutes a current, and as the successive rushes are in opposite directions, the conducting rods become the seat of an alternating current of high frequency. Therefore in the space surrounding the bent rods there will be set up an alternating magnetic field. The magnetic field surrounding the wire will be strongest when the rushing current is strongest, and this occurs precisely at the moment when the condenser has been emptied and before it is charged up again by the continuance of the rush. But the electric field, which will also be alternating, and which lies in the crevice between the plates, will be strongest when the rush either way is finished, and the charge in the condenser is at its height. So we here see that the

* Read before Electrical Club, University of Toronto, Nov. 1910.

magnetic and electric fields occur at different places and have their maxima at different times. It has been experimentally proved that under such conditions no electric waves can be emitted; it here being taken for granted that electric oscillations occurring under the proper conditions give rise to what are known as "electric waves," a phenomenon which many years ago has been well studied and described by many eminent scientists.

Let the condenser of Fig. 1 be opened up into the form shown in Fig. 2, with the two metal plates, .WW., extended

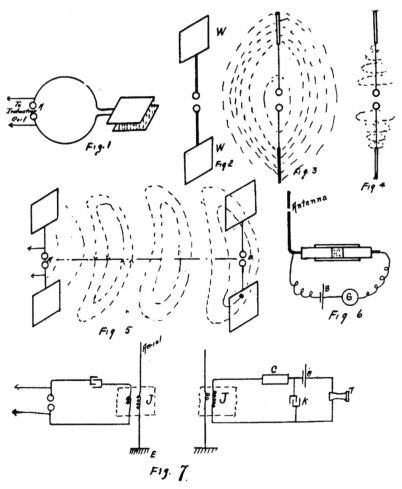

out like wings, and conductor with spark gap, A, in the middle made straight. As before, air is the dielectric. When they are charged, there will be an electric field from one to the other as in Fig. 3. If a current were rushing up and down from one to the other, there would be a magnetic field, as shown in Fig. 4. Now, when a spark is made to pass, setting up a series of oscillations, the electric field between the wings will extend, as the dotted lines show, right across the space where the magnetic field will occur during the rushes of the current up and down the rod. The electric field will not have died away before the magnetic field has begun to grow, so that both kinds of fields can be present in the same space at the same time, and while the electric lines will be mainly parallel to the conductor the magnetic lines will be transverse whirls around it. For the production of electric waves it is necessary that there should be both an electric field and a magnetic field at the same point of space and at the same time. This simple apparatus is capable, then, of throwing off into the space surrounding it an electric wave at each oscillation. Such waves do not return back into the system, but go travelling off into space with the speed of light, following one another as in Fig. 5. The more the metal plates present of free surface, the more freely do they radiate off electric waves. If they consist of flat metal the electric waves radiate off more freely from the flat faces than from their ends or edges. If set vertical they radiate out mainly in front and behind, but they may be set horizontally at the ends of the system. The late Professor Hertz devised this simple apparatus in 1887 for the manufacture of electric waves. It is known as a Hertz Oscillator.

Now, to detect these waves, Hertz placed in another part of the room an exactly similar apparatus with two plates and a minute spark gap, as shown on the right of Fig. 5. The spark gap B was not more than 1-1000 inch, otherwise the induced electromotive force was not able to make the spark jump across the air space. Hertz, on placing an oscillator and a detector like these, a few feet apart in a room, found that a spark occurred at B every time a spark occurred at A. This induced emf. was seen to be due to waves emitted from the oscillator and caught by the detector. Hertz was also able to show that these electric waves could be reflected from a large sheet of metal or collected by a parabolic mirror or refracted through a prism of pitch; in fact, they behaved like waves of light, only they were quite invisible.

These results obtained by Hertz led other men to experiment. Branly discovered that a heap of metal filings possesses the very curious property that although they are usually a very bad conductor because of the innumerable imperfect contacts among the particles, they yet become a much better conductor when an electric spark is made anywhere near them. Lodge found that the imperfect contact made a pointed wire resting against a metal plate to be sensitive to the action of an electric wave. Such imperfect conductors are called coherers. Taking advantage of Branly's observations, Lodge constructed a coherer of metal filings enclosed in a small glass tube between metal rods inserted at the ends, and this proved very sensitive, but he found that it required to be tapped after each operation to cause the filings to decohere, and proposed methods of automatic tapping which will be touched on later.

The method in which the coherer was used to detect the electric signals is as follows: The coherer was connected to a voltaic cell and a galvanometer in series. The coherer has so poor a conductivity that practically no current flows. Directly an electric wave falls upon the coherer or upon the wires attached to it, it sets up oscillations, and probably also sets up minute sparks in the air gaps between the filings. Whether this is so or not, it increases the conductivity of the filings so that the cell is able to send enough current through the circuit as is detected by the galvanometer (see Fig 6.) Lodge found considerable advantage in fastening to the coherer an extended wire as an antenna to receive the wave. In his early experiments he put the whole apparatus in a tight metal box to exclude the effects of any stray waves, the only thing projecting out of the box being the receiving antenna. With such a coherer he was able to extend the distance between transmitter and receiver to two miles.

He also extended the distance by using at the spark gap a single polished ball between two smaller ones.

In the Hertz upright oscillator, as in Fig. 5, the capacity plates at the top and bottom ends of the oscillator are the places which become most highly charged, the middle part of the connecting system being as it were a node in the waves. The same would be true of the same apparatus if used as a detector. The ends are the places which receive the maximum charges of potential. The middle part is the seat of the strongest currents, and whole loops of waves are sent out. But this is no longer the case if the oscillator or receiver is earthed, as to the lower half, and the node in the electric wave occurs just above the earth and waves of which half are in the air are sent out.

This idea of having half the wave travelling in air and the other half in the ocean or earth is of great advantage. The waves maintain continuous contact with earth, are not sent through space, but undulate along the surface, and cannot be reflected by earth into space as they would be if the waves were closed upon themselves; so this prevents the dispersion of energy in any disadvantageous direction, and at the same time facilitates its transmission in the useful direction, that is to say, in the horizontal, and in practice it is found that the intensity of the wave does not diminish more rapidly than the simple inverse of the distance from the centre of emission.

Since the earth takes an essential part in the transmission of waves it must be of good conductivity: wet soil or sea. Earthed apparatus does not work over a dry desert. The movement of the half loops of electric waves outward from an earthed oscillator as distinguished from the non-earthed oscillator used by Hertz and Lodge, is hindered by a bad conductivity on the earth's surface, but is helped by the fairly good conductivity of the sea. On the other hand, for non-earthed oscillators emitting whole loops of waves, the conducting power of the earth is no help.

The first wireless telegraphing was done by Sir Oliver Lodge, who, in 1894, sent wireless signals through stone walls and from one building to another, using as a detector a filings coherer as described, the system being untuned. Signor Marconi at first made his experiments with a coherer put to earth in circuit with a single cell and a sensitive telegraphic relay, which in turn actuated an ordinary Morse instrument, printing dots and dashes. In 1898 Marconi adopted the plan now claimed essential to his system of using a tall mast like a lightning conductor, the lower side of the spark gap being connected to earth and called "the aerial wire," or "antenna," or "earthed vertical oscillator."

The frequency of the oscillations is determined by the capacity and self-induction of the apparatus and also by the length from the node to the ends and the surface presented. Also any evils introduced into it will increase its self-induction and lengthen the waves. An oscillator like Lodge's or Hertz's has relatively large surface and small self-induction and will radiate so freely that the wave train will die out after very few oscillations. Owing to the fact that the Marconi aerial resembles the Hertz oscillator in possessing a large radiation surface, it cannot send out such trains of waves as would be required for syntonic working, or in other words, it is necessary to tune the instruments and the aerial. In none of the early operations could tuning be secured, and they were liable to interference from stray disturbances in the atmosphere.

With regard to tuning we must observe that the receiving apparatus is a vibrator similar to the sender, and therefore on the first arrival of an electric impulse electrical vibrations will be produced in it, having periods similar to that of the receiving apparatus. The same effect will be produced with the arrival of each wave, and since the effects are additive, it is necessary that the arriving waves should have the same rhythm or period as those which are set up in the sky rods, otherwise the vibrations set up by the successive impulses will overlap each other, irregularly weakening each other, and producing what is called interference of waves. Therefore when waves are striking the rceiving apparatus it is the duty of the operator to vary the capacity and inductance of his apparatus so that his apparatus will vibrate with the same rhythm as the distant transmitting apparatus, and in practice the high clear note is found to be the one desired, that is, if a telephone were connected in the coherer circuit one should hear clear high notes for best working. When the receiving apparatus is so set as to receive messages it will generally be found that a hum can be heard in the telephone. Stray disturbances in the atmosphere, perhaps static disturbances, or stray waves emitted from some place or other cause this, and therefore if station A wishes to converse with station B the apparatus of the receiving station must be tuned such that the required message is heard above these other minor ones. As stated before, the high note is required for good working, and the higher the note the easier it is for the operator to distinguish it. Also in this connection it might be well to note that waves should not be damped. Damped waves diminish and die away after a few vibrations. Damping may be caused by too high resistance in conductors, spark gaps, etc. A high sky rod reduces damping and so does putting part of oscillator to earth. If a source of persistent oscillations could be applied, the receiving apparatus could be tuned to the sending apparatus, and thus the whole system be rendered far more sensitive for long-distance work. To do this the sending apparatus must in some way be linked up with an oscillation circuit, and the distant receiving apparatus should similarly be supplied with an oscillation circuit; then the two can be tuned to the same frequency of oscillation. There are more ways than one of associating the wave apparatus with an oscillation circuit. The most usual one is by means of an oscillation transformer, called in telegraphic language a "jigger," and is due to Sir Oliver Lodge. This induction transformer is a simple arrangement for a primary coil of one or two turns surrounding or surrounded by a secondary coil of a larger number of turns. A simple form of jigger is shown in Fig. 7; J is the jigger, represented by two separate spirals, K the condenser, C the coherer. When the primary coil is traversed by oscillations they induce other oscillations in the secondary circuit, but these last are feeble unless a condenser is put across the secondary terminals and the two circuits brought into resonance.

Besides the advantage gained by tuning, there is another advantage of putting the oscillation transformer into the receiving circuit. For at the node of the receiving aerial the potential is a minimum, and the current a maximum. Therefore as the coherer depends on the potential and not on the current, the node is a bad place to insert the coherer, as was done at first. By placing in the node the primary of the transformer and inserting the coherer in the secondary, the increased voltage so applied to the coherer was found to increase its sensitiveness considerably, and therefore greatly extended the range of working. Marconi being desirous of working over still greater distance, adopted (in 1899) the jigger into his arrangements, and was able to signal eighty-five miles. For transmitting oscillations the transformer was generally constructed as follows: It consisted of a square wooden frame wound over with a number of lengths of highly insulated, stranded copper cable, joined in parallel, so as to make a primary of one turn. Over this is wound a secondary of from five to ten turns. The oscillation transformer is usually immersed in oil.

Having obtained the syntonisation of the stations in such a manner that the receiving station will only answer to the wave emitted by a station tuned to its period, it will be easily understood that if there be several receiving stations, each one tuned to a different period of electrical oscillations, and whose corresponding length of wave is known at the sending station, this latter can tune its own apparatus in such a manner that the despatch shall be received by one determined station among these, and then modify the tuning in order to communicate with another, the message with the first having ceased, and so on.

The problem of multiple communication, that is to say, the communication simultaneously to several stations in any direction and in a similar radius of action, is therefore intimately associated with that of syntonisation. It will be easily understood that a given station will have greater importance even from the business point of view in proportion as the number of special stations with which it can communicate, be greater.

Marconi has contrived a way by which he can at one and the same time transmit from a single station several messages. In 1903, in Italy, he sent five messages at once from a single sky rod. As spoken of before, tuning can be accomplished by changing capacity, inductivity or both. Signor Marconi, for greater facility and certainty of action, causes conductors of different capacity and inductance to be thrown into contact with the aerial and having it arranged so that this can be done for each transmitting apparatus, then no two will have the same wave form and thus two or more signals can be leaving the aerial at the same time. The receivers can be similarly fixed so that each will take care of its own messages.

In passing from the theoretical discussion to a discussion of practical apparatus as is used by wireless companies, it might be stated that to fully discuss the theory of radio-telegraphy, higher mathematics should be made use of, as, for instance, to show relations existing between wave length and capacity. Also the formulae for induction frequency, wave length and capacity could be derived, but space is too limited for that phase of the subject.

An ordinary wireless station contains at least one transmitting set and one receiving set. If it is a station that is supposed to work within a radius of five hundred miles, the transmitting set will likely consist of a motor generator set, an oscillation transformer, a variable inductance and capacity in the form of coils and condensers, a spark, gap, a key, and an aerial. The receiving apparatus has the same aerial as part of it, a detector of some kind, as, for example, the coherer, a tuning device consisting of inductance coils and condensers, a telephone or some recording telegraphic device or sounder, including the necessary wiring and switches. A large transmitting station, such as a trans-Atlantic one, needs more apparatus than a simple motor generator set for generation of power. In such a station we find the accumulator house, the condenser house, a boiler room, engine room, an operating room and a high tension room and an arial covering several acres, all of which will be discussed in detail.

Transmitting Apparatus.

For short-distance spark telegraphy the necessary high potential is always obtained by the use of an induction coil or transformer; either a single instrument or a number of induction coils or transformers may be employed, having their secondary circuits joined in series and their primary in parallel.

The most usual appliance is an induction coil of the ordinary type, having a spark length of ten inches and taking current from storage batteries or from a small single phase alternator. The induction coil is placed on the table in the transmitting station, or it may be fastened to the wall. It may be operated either by alternating current or an interrupted continuous current.

The next element in the transmitting apparatus is the signalling key for interrupting the primary circuit in accordance with the signals of the Morse alphabet or Continental code. This must be a quick break key with a long ebonite handle easily operated, and has generally a magnetic blow-out in connection with the platinum terminals between which the interruption takes place. Marconi has tried several kinds of keys. Some have the points touch in air, some in a recipient filled with paraffin. He uses in his key circuit a condenser in shunt so as to do away with self-induction sparks on opening or closing the circuit. Also if an alternator is used a provision is made to save it, since it is running all the time. Marconi puts in parallel with the key a reaction coil whose reactance when its core is completely immersed is such as to damp entirely the exciting current. Every time the key is closed the current excites the primary of the transformer and discharges are produced and on the other hand when the key is raised, the exciting current is annulled by the reactance and the discharges cease without any necessity of altering the working of the alternator.

Reference has previously been made to the oscillating transformer or jigger. Marconi prefers to connect the aerial inductively with the energy-storing circuit by the oscillation transformer. It consists of a few turns of primary winding wound on a wooden frame and more turns of a secondary connected in series with the aerial that has a variable inductance between this secondary winding and the earth connection. This last mentioned inductance is really a turning coil. A condenser is put in series with the line leading to one end of the primary of the jigger and the spark gap is connected in parallel with the primary coil having this condenser in a line between the spark gap and primary winding.

Reference might here be made to what happens. when two oscillation circuits are connected together inductively and in tune with one another: oscillations set up in one circuit results in the production of oscillations in both circuits, having two frequencies, one greater and the other less than the natural frequency of each circuit when separate. Hence when employing an induction coupled antenna which has been syntonised with the condenser circuit, it is necessary to bear in mind that oscillations of two frequencies are set up in the antenna and waves of two wave lengths radiated from it, one greater and the other less than the wave length corresponding to the natural frequency of the antenna taken alone. One of these waves has greater amplitude than the other, the longest wave length being the least damped and therefore having the greatest integral value. The wave lengths approximate to one another in proportion as the coupling is made weaker but then they also diminish in amplitude so that by the employment of a weak coupling, which can be done by separating the primary and secondary of this transmitting jigger, we get a radiation which is a little feebler but of a single wave length, whereas by close coupling we get more powerful waves but waves of two wave lengths radiated, and the receiving antenna must accordingly be syntonised to one or other of these wave lengths.

As regards condensers for the oscillation circuit, although a Leyden jar is a bulky form of condenser in comparison with its energy storing power, nevertheless its simplicity still recommends it. The main condenser consists of a battery of them joined partly in parallel and partly in series. It is very important that the capacity of the condenser is exactly known and the jars selected so as to be exactly equal to eliminate as far as possible electric brush discharges. The condenser can be constructed of plates immersed in oil. Brush discharges are thereby prevented and accuracy of tuning is secured by preserving a constant known capacity in the condenser circuit. Exact syntonisation between condenser circuit and aerial circuit is absolutely necessary. Other things being equal the radiation will be in proportion to the mean square value of the current flowing into the base of the antenna. The current may be measured by inserting in that point a hot wire ammeter.

Another important element in the transmitting arrangement is the spark discharge. When large capacities are being employed the noise of this spark is distressing and any one who understands the code can read the messages at a great distance. The spark balls should be enclosed and have alkaline material close by for absorbing acid vapors or gases. It is of great advantage to blow a jet of air upon the spark gap to quench the arc. A spark gap can be done away with and a Poulsen arc substituted in its place. This is a device invented (in 1903) by Poulsen of Copenhagen. He produced an electric arc between a carbon rod as a negative and a copper rod as a positive terminal, the latter being kept cool by water circulation. This apparatus gives very powerful undamped oscillations, the frequency of which by a proper selection of capacity and inductance, can be made to be as high as a million or more, and quite within the range of those required for radio-telegraphic work.

Marconi in his large stations has made use of a new device for discharging, known as the High Speed Discharger. In one form of it, there are caused to rotate rapidly by electric motors or other means two discs with perforations. Between these, and insulated from them, another disc (with its plane at right angles to the other two) rotates at a high speed. This plane and the two discs are connected in the circuit with inductances and condensers. At a certain potential, the condensers discharge with oscillations across one or other of the air gaps between the rapidly revolving disc. The arc discharge which attempts to follow in the track of the oscillation is however prevented by the rotation of the discs from taking place. A very high piercing note is produced by this sparking apparatus.

The length of antenna varies with the size of the station. A small five kilowatt station for operating up to 500 miles would have a tower about 150 feet high, situated on some high point, and having bare wires leading in a downward direction. Another scheme is to use two towers joined by bare wires and a tie-arrangement leading to receiver. In the very large stations as for example the Glace Bay Station, the Clifden Station on the coast of Ireland, or the Poldu station of Cornwall, more extensive aerial is required. At the Glace Bay Station at Cape Breton Marconi has four towers built with a cone shaped envelope of wires suspended from the top of the towers, thus making it possible to have a large radiation surface. The aerial apparatus at the German station of Nauen near Berlin is pretty extensive. That company uses there a steel tower 328 feet high built so as not to vibrate, the tower being triangular and thirteen feet to a side. The tower is insulated from the earth as well as the guy cables. The antenna is sloped like an umbrella, the tower representing the rod. The aerial is divided into six sections, the upper part consisting of six sections of 54 wire cables, nine in each of the six sections. At a point 82 feet from the top of the tower each cable divides into three so that there are 162 cables altogether. The aggregate surface enclosed is fifteen acres. The cables are insulated from the ground and led in parallel to the station. The ground contact is made through a system of 108 wires buried in low wet ground. These wires divide as they diverge, and the area enveloped is thirty-one acres. There are altogether 324 wires in the ground. Most of Marconi trans-Atlantic stations are situated near the coast in the very wettest piece of ground he can find in the region. The station at Glace Bay is in a very desolate, lonely swampy place.

Just here it might be well to speak of the transmitting apparatus of the Nauen station. The operating rooms occupy a brick building, the ground floor of which contains the dynamo. the transmitting office and the work ship. All the high tension apparatus is placed in the second storey where it is free from dampness and hence more perfectly insulated. The power is furnished by a 35 H.P. engine which drives a single phase alternator coupled to an exciter on the same shaft. This generator at a speed of 750 R.P.M., furnishes 25 kilowatts of electrical power in the form of a single phase current of 50 cycles per second. In the circuit are fusible plugs, voltmeter, ammeter, frequency indicator, a transmitting and cut out relay. Four large inductance coils are in the dynamo circuit, in addition to the four transformers which produce the high tension transmitting current. The transmitting circuit also includes a battery of

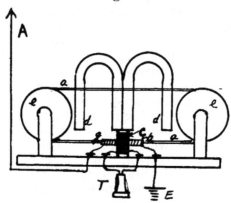

Fig 8.
MAGNETIC DETECTOR

360 Leyden jars arranged three in series by 120 abreast and having an aggregate capacity of 400,000 coulombs.

This station just described has sent messages 1500 miles over water and received telephonic messages from St. Petersburg a distance of 840 miles.

Marconi, to transmit messages across the Atlantic, uses an installation of 75 kilowatt capacity, the voltage being transformed up to about a quarter to half a million volts.

Receiving Apparatus.

The discussion of receiving apparatus centres around that of a special part of it known as the detector. There are indeed many forms of detector of electro-magnet waves from the old filings coherer to the latest inventions of magnetic and electrolytic detector and lastly the famous Fleming Valve Receiver.

Marconi first used a coherer as a detector. It had nickel filings with four per cent. silver in it. The dimensions of such a coherer are 1½ inch long, 1-10 inch internal bore with silver stoppers 1-5 inch long and the space between for filings being 1-50 inch. With this delicate apparatus 1.5 volts was the maximum voltage he used and one milli ampere was the maximum current sent through by a single cell. He used a decoherer in the form of a hammer actuated by an electromagnet which just struck a rubber pad on the coherer tube.

The second type of coherer used was a magnetic one. It was discovered that electric waves can exercise a demagnetizing influence upon a highly magnetized small steel needle and a magnetic detector of waves on this principle was made. Marconi for telegraphic work invented an improved form of detector a diagram of which is given in fig. 8, made as follows: Two wooden discs e,e, grooved on the edges are driven mechanically by clock work. An endless band a,r, made of a bundle of fine silk-covered iron wires, is arranged like a belt over these pulleys and moves forward at the rate of about three inches per second. At one place the iron band passes through a glass tube, g.b., on which is wound a coil of insulated wire through which oscillations can be passed and this coil is embraced in the centre of another coil, c., connected with the telephone T. A. pair of horse-shoe magnets are placed with their similar poles together opposite to the last mentioned coil as shown in the diagram. If electric oscillations pass through the coil wound round the band they change the magnetic state of the iron and generate an induced in the secondary coil and hence a current and sound in the telephone. The operation of this instrument was considered by Marconi to be due to the power of electric oscillations passing through the coil surrounding magnetized iron to annul the hysteresis of the iron. A magnetic detector like this is found to work, when there are static disturbances a little better than some other detectors.

The third and present detector used by Marconi is known as the Fleming Valve Receiver, invented by J. A. Fleming, M.A., D.Sc., Pender Professor of Electrical Engineering in the University of London. He first used an ordinary incandescent lamp with carbon filament and having a metal plate included in the glass bulb or a metal cylinder placed around the filament, the said plate or cylinder being attached to an independent insulated platinum wire T, sealed through the glass. When the carbon is rendered incandescent by electric current, the space between the filament and th plate, occupied by highly rarefied gas possesses a unilateral conductivity and negative electricity will pass from the filament to the plate but not in the opposite direction. This depends upon the well-known fact that carbon in a state of high incandesceuce liberates negative ions. A practical lamp has been made—a small incandescent lamp with a rather thick filament taking about 2 amperes at a terminal voltage of 12 volts. *Fig.* 9 shows how this detector is used in radio-telegraphy work. The antenna A, is coupled through an oscillation transformer with a circuit which includes the valve, O, and the fine wire coil of an ordinary 10-inch spark induction coil, I, the low resistance coil of which is in circuit with the telephone T. Condensers are placed across the secondary circuit of the oscillation transformer and also in series with the fine wire coil of the large induction coil. By suitable adjustments of the capacity of this condenser the circuits are brought into resonance. Oscillations taking place in the antenna, due to the impact of electric waves upon them are then transformed by the jigger, rectified by the oscillation valve and sent through the fine wire coil of the large induction coil in the form of a unidirectional but intermittent current and these oscillations are again transformed

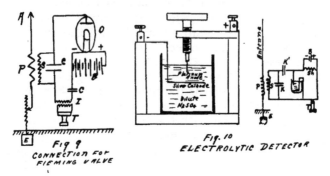

Fig 9
CONNECTION FOR
FLEMING VALVE

Fig. 10
ELECTROLYTIC DETECTOR

up in current value by the large induction coil. In this way a sound is created in the telephone. So used, the oscillation valve becomes one of the best long distance receivers for electric waves yet devised .

At the top of the departmental store of John Eaton, Esq., Toronto, is a wireless station which has all three of the Marconi type of detector installed and in first class condition, the station being of two kilowatt capacity. Power is furnished by a motor-generator set at 500 volts and the voltage is transformed to 12000 by the oscillation transformer. The spark used is five-eights inch long, being two gaps in series. The operator can intercept messages sent out from the Atlantic stations near New York and is in daily communication with the Clarke Great Lake stations. The high efficiency of this station is due to the excellent tuning apparatus a wiring diagram of which is shown in *Fig.* 11. The Fleming valve is the detector shown. Two valves are installed, only one being in use at a time. Neverthless they find the magnetic detector the better during static disturbances in the atmosphere. A fine adjustment for tuning is to be had in the billi-condenser. When the operator is just listening the switch is on the stand-by side but when communication is opened up with any one particular place it is thrown over to the tuning side and the receiving apparatus is tuned to read the required message, the tuning cutting out all stray disturbances and messages of other stations. The wave length for this outfit is about 300 metres. It possesses an advantage over the Clarke system in the operator being able to receive and transmit at practically the same instant: all that is necessary to receive is to stop working the key whereas in the Clarke Lake stations a lever has to be moved which (to accomplish) takes a little time.

Besides the form of detector used by Marconi an American company headed by Fessenden uses an electrolytic detector. It is shown in *Fig.* 10. It consists essentially of a vessel having as

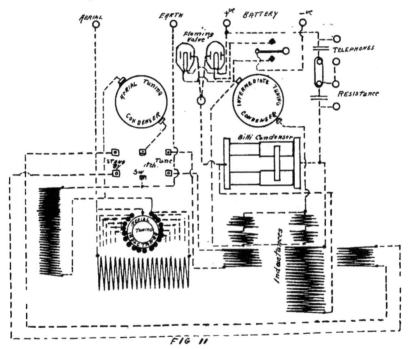

DIAGRAM OF CONNECTION OF A MARCONI
RECEIVING APPARATUS USING FLEMING VALVE

one electrode a very fine short wire of platinum offering therefore an extremely small surface. This electrode is generally made the anode. The other electrode is a silver plate. The electrolyte is sulphuric acid. The cell becomes polarized but electric oscillations depolarize it, thus allowing a current to circulate and giving rise to a sound in the telephone. When the wave ceases it is again polarized. It is quite automatic in action and very efficient.

Effect of Atmospheric Conditions and Daylight.

The first attempt to conduct radio-telegraphy extending over many hundreds of miles revealed the important influence that atmospheric conditions have upon such telegraphy, especially the effect of sunlight upon it.

Marconi conducted experiments on board S. S. Philadelphia on the Atlantic. He noticed that it was possible to receive signals by night when they could not be detected by day. He found that there was very little difference in the signals received by day and by night until he was 500 miles from Poldu. After that, day signals began to weaken and at 1500 miles were hardly perceptible. At daybreak the effect was worse. This is attributed to the fact that light dissipates negative charges of electricity. Some scientists say that the atmosphere especially under the influence of daylight is in a state of ionization and it has been shown that these point charges of negative and positive electricity are set in motion by long electric waves travelling through space and they therefore partially absorb the wave energy.

On account of this effect of daylight messages are sent mostly at night, Press messages being sent from 10 p.m. to 1 a.m.

Another important fact could be mentioned in connection with long distance work and that is in the small degree to which the curvature of the earth seems to affect intercommunication between stations employing earth-connected antenna. It is well known that rays of light and sound are diffracted to some extent around obstacles but the long Hertzian waves from an earthed antenna appear to pass round a one-eighth part of the circumference of the earth without extravagant diminution of amplitude other than that due to distance and atmospheric absorption. The possible cause of the advantageous transmission round the terrestrial sphere is due to the earth connection both of the transmitting and receiving antennae, whereby both these antennae and the earth are practically converted into a single oscillator.

Radio-telegraphy like any other branch of electro-technics has its unsolved problems and it can not be said that a complete explanation of the nature of the propagation of electro-magnetic waves over and round the surface of our globe has been reached which is beyond dispute.

CHICAGO'S SHIP AND DRAINAGE CANAL.*

The Engineering Society of the University of Toronto were afforded a highly instructive address and a rare opportunity on December 12th, when Mr. Isham Randolph, C.E., delivered an illustrated address on Chicago's Ship and Sanitary Canal. While the address was exceedingly comprehensive, it was evident, as Mr. Randolph said, that so vast an enterprise could not be justly dealt with in one address. In fact, Mr. Randolph said the subject could be treated to more advantage under several heads, as, metropolitan, state, interstate, national, and international. In the course of his address Mr. Randolph said Chicago is but 74 years old and the great problems arising from sudden growth may be well imagined. He first took charge of Chicago's work in 1893, a few months after the great fire. At that time Chicago was a straggling sort of a city of about 400,000 population. Streets, in many cases, were below water level an dtremendous operations have taken place to put them in their present condition. Lake Michigan has always been Chicago's source of drinking water. The first lake tunnel ran two miles into the lake, but this was later regarded as insufficient and it was run 2 miles further out. At present there are several running out 4 miles from shore. In 1886 there was a commission appointed concerning this matter of sewage. This commission considered three propositions, (1) an intercepting sewer for conducting sewage to southern part of the lake, (2) a settling basin land filtration plant, but as this affected territory belonging to Indiana that state raised objection to this, (3) carrying the sewage douw the Illinois valley. But with the failure of these propositions the present problem arose. In 1889 the Legislature of Illinois passed a sanitary law, which materially affected Chicago's problem. Whereas 20,000 cubic feet of water per minute was obtainable, they planned a channel for obtaining 600,000 cubic feet per minute.

This canal was to be 200 feet wide on the bottom and 18 feet deep, but in reality it was 202 feet wide at bottom and 22 feet maximum depth. Through the rock the channel was 160 feet wide at bottom and minimum depth was 22 feet. In the rock there was a slope of 2¾ inches per mile and in the clay and sod a slope of about 1½ inches per mile. The cut was made through the rock with a channelling machine and was taken out in 3 stokes with widths at the bottom of 160 feet, 161 feet and 162 feet respectively, these offsets being allowed for the proper manipulation of t he channelling machine. Mr. Randolph was connected with this work for over fourteen years and it assumed enormous proportions during that time. One might reasonably ask whether this enormous expenditure was merited, for the sanitary district of Chicago has an area of only 356 square miles

---

* Courtesy of "The Canadian Engineer," Toronto.

.64 of 1 per cent. of the area of the State of Illinois. This comparatively litle area expended $63,000,000 for his prodigious undertaking. The report of the health commissioner of Chicago will perhaps tell whether this really was a paying proposition. The report says that in 1891 Chicago had about the highest death rate from typhoid of any city. The result of these sewage improvements was that from 64.1 per 100,000 population, the death rate in Chicago for typhoid decreased to 23.5 per 100,000 population. And from having the highest death rate of any city it now has about the lowest, having only 15.6 per 1000,000 death rate. According to the report of the health commissioner, 11,148 typhoid deaths had been prevented in 9 years by this system. Mr. Randolph, in his address, expressed the wish that Toronto had such a drainage canal.

The Chicago ship canal is eventually 200 feet wide, 26 ft. deep in mid-stream and 18 ft. deep at the banks, and some 28 miles long. Mr. Randolph has many exceedingly interesting slides of the works, a remarkable feature being the description of the many various appliances used and in some cases devised for the first time to facilitate in the excavating and dredging on this work. Among some of the most interesting slides were those of the Brown level conveyer. which would often complete its operation in 52 seconds in handling 10,000 pounds. At Lockport there are seven gates, interesting from size. They fill gaps of 32 feet, are 20 feet high, have a weight of 62.000 pounds. Two men can raise and lower them in 15-foot pressure of water. The "Bear trap" dam at Lockport is also a remarkable feature—it is 160 feet on the crest, has 12 feet oscillation and the hinges are anchored 20 feet. The placing of this interesting type of structure at Lockport met with much opposition from commissioners and engineers alike and they predicted it would be a failure. It is, however, a decided success. There is. also on this canal the so-called "Butterfly" dam, the only one of its kind in the world. It is 184 feet long, 34 feet high and has a pressure on the bottom, when closed, of 3,776,000 pounds. Mr. Randolph solved the hard problems of successful dredging by using the hydraulic dredge. This wonderful machine worked so well that another was built on the works. With the hydraulic dredge the contractors were able to remove 1,400,000 cubic yards of material at a cost of about 5 cents a yard. The Mississippi Navigation Commission following the example of Mr. Randolph has since employed the hydraulic dredge and this is making the navigation of the Mississippi possible. The dredge as used on the Mississippi cost $150,000 and 1,500 yards an hour can be removed with it. The cableway, an invention which was conceived on this Chicago's work. is another mechanical aid that has helped the wor kof excavation in engineering. The aerial dump has done marvels since its invention. Such appliances and the channelling machine, which on the Chicago work cut a gash of ⅜ inch and did 100 superficial feet a day, have made the work of excavation less tedious, and in some cases, at all possible.

In the Panama they are far exceeding now work done on the Chicago canal. This year they have excavated 135 million cubic yards. They excavated only 66 million last year and only 37 million cubic yards the year before last, so a tremendous progress is shown in the speed of excavating. Bridge work over the Chicago River has been very extensive. Eleven of the rolling lift type have been built across the Chicaga and more will be built. The principle of this very efficient bridge is that as the channel span opens the approach span drops. This type has been found very satisfactory.

Mr. Randolph's lecture was listened to with great interest by student and instructor alike, and a hearty vote of thanks extended him by the Society.

THE ENGINEERS' CLUB OF TORONTO.

C. R. Young, B.A.Sc.

For some years past various unsuccessful efforts have been made to establish in Toronto a club offering to engineers, architects, surveyors, chemists, and others engaged in applied science pursuits, all the privileges of the usual social club with technical advantages as well. It has been the dream of almost every member of the original Engineers' Club, incorporated under the Benevolent Societies' Act in 1902, that sometime and somehow the Club should expand into an institution occupying and possibly owning a club house of its own, of sufficient capacity to accommodate within its walls the various technical organizations existing and holding meetings in Toronto. In spite of the many sincere and earnest attempts to bring about this extension, the monetary difficulties in the way proved too much for those who first interested themselves in the commendable undertaking.

Finally, in the early part of the year 1910, a movement was set on foot aiming at the establishment of a "Social Club for Technically Trained Men." The final name was left to be chosen later.

It was recognized that with a membership drawn from engineers and surveyors alone who, up to that time had composed the Engineers' Club, it would be impossible to maintain an advanced Engineers' Club, and that the co-operation of architects, industrial and assaying chemists and all technically trained men working along applied science lines, as well as business men of allied pursuits, must be secured.

It was the intention of the promoters from the first that

should the required support be pledged, at the proper time the existing Engineers' Club should be approached, and the two bodies united under a common name, and on a basis equitable to all, and with the strictest justice to each and every member of the original club.

In its early stages the proposition met with much success, and in the course of some four months the names of 360 men, including members of the existing Engineers' Club were secured, who agreed to join the new organization. At this stage, however, an insurmountable difficulty developed—that of securing a charter suitable for such a Club. Finally after much consideration, the organization committee of the "Social Club for Technically Trained" men informed the Engineers' Club that a charter carrying with it the usual club privileges would be more likely granted were it sought, not by the Social Club, but by the existing Engineers' Club, and with this the former organization ceased activity.

Accordingly, at a regularly-called meeting on August 5th., the Engineers' Club of Toronto appointed seven members to apply for a charter under the Ontario Companies' Act, and to follow out, if possible, the procedure essential to the establishment of a Club on the proper basis. It was tacitly agreed that the routine of the Club's business should in the meantime, be discharged by the old Executive Committee.

The desired charter was granted, after much unavoidable postponement, on September 27th., and the following seven members were named therein as provisional directors: A. B. Barry, C. M. Canniff, Willis Chipman, John Galbraith, J. G. Sing, J. B. Tyrrell, and A. J. Van Nostrand. On November 15th, the provisional directors enacted, under seal, after careful consideration and many amendments, a code of new bylaws suitable for the prospective wider life and extended membership of the Club, basing them upon the draft of the "Social Club for Technically Trained Men." These By-laws were finally adopted, without further alteration at a duly called meeting of the Club on December 1st., and the re-organized Club at last began its existence.

At this meeting the directorate was increased from seven to fifteen, the additional eight directors being drawn from the membership of the original Engineer's Club for obvious reasons. The full list thus became as follows: R. A. Baldwin, S. P. Biggs, W. A. Bucke, C. M. Canniff, Willis Chipman, John Galbraith, W. A. Hare, C. H. Heys, E. A. James, J. G. Sing, C. B. Smith, L. J Street, J. B. Tyrrell, A. J. Van Nostrand, and C. R. Young. These directors hold office till February 2nd, 1911, when the first regular annual election will take place, and when it will be possible to secure full representation on the Board, of all the interests concerned.

The provisions of the By-laws respecting membership will,

no doubt, be of interest. It is stipulated that the Club shall be composed of engineers, architects, surveyors, industrial chemists, and others, who may be connected with or interested in engineering or allied pursuits. The membership of the Club is limited to 800, composed of 500 resident, 200 non-resident, 25 life, and 75 associate members. Ordinary resident members pay an entrance fee of $50, and an annual fee of $30, while for non-resident members the corresponding fees are $25 and $20. The annual fee is reduced in the two cases to $25 and $15 respectively for prompt payment. Each member must be at least 25 years of age.

The members of the original Engineer's Club are given preferential treatment to the extent of the remission of the entrance fee, since assets of considerable value were automatically transferred to the re-organized Club, on its incorporation under the Ontario Companies' Act. Those members of the old Club who, for any reason, do not wish to take full advantage of the extended privileges offered by the new Club, are, on payment of an annual fee of $5, entitled to the use of the Library, reading room, and lecture hall. The improvements which will be effected in these privileges will considerably enhance the benefits enjoyed by the members of the old Club.

Very soon after the adoption of the By-laws by the members of the Club, a Directors' meeting was held at which officers and members of standing committees were elected and other important business was transacted. The officers elected to hold office till February 2nd, 1911, were: President, C. M. Canniff; First Vice-President, Willis Chipman; Second Vice-President, A. J. Van Nostrand; Treasurer, L. J. Street. R. B. Wolsey was appointed permanent secretary of the club.

It was also decided to engage under a lease renewable on the present terms, for a period not exceeding two years the rooms on the second and third floors of 90-98 King Street West, including the rooms occupied in the past by the old Engineer's Club and the Ontario Association of Architects. These rooms are now being decorated and furnished in a manner suitable for the requirements of the new Club.

The advantage offered by the re-organized Engineer's Club cannot fail to prove highly attractive to technically trained men. With a down-town social club there will be an opportunity for engineers, architects, surveyors, chemists, and those whose pursuits bring them into contact with technically trained men, to meet and become better acquainted. Here members will be in a position to suitably entertain their business friends and clients at luncheon, diner, or otherwise; and out-of-town members may meet their city associates and clients and transact their business. To this end, it is proposed to provide, in conformity with a suitable standard of comfort and elegance, a Grill Room, Smoking Room, Sitting Rooms, Billiard and Card Rooms, and Private

Conference Rooms. A further extension of privileges will be afforded by affiliating with engineering clubs of other cities, and in return for the hospitality extended to Toronto engineers, visiting members of the engineering and allied professions will be entertained here in a befitting manner. Dormitories are being provided to accommodate such visitors as may wish to lodge in the club house during their stay in the city.

While the social privileges of the new club have been first mentioned, the technical side of its activity will by no means be neglected. A reading room, properly isolated in a quiet part of the building will be amply furnished with technical and other journals and magazines, and a first-class technical library will be gradually built up. In this work the various technical societies, meeting in Toronto, will co-operate with a view to economizing resources and avoiding duplication. The lecture room will be used by these organizations for their meetings, and the Engineer's Club itself will provide a technical programme during the winter months. Papers read at its meetings will probably be printed later in the form of proceedings.

A proposal which will, no doubt, be carried into effect, and which will enhance the usefulness of the club, is the establishment of a bureau of information concerning technical work to be done, and men fitted by training to do such work.

A club affording privileges of this character, and free from politics or other clique-forming influences will undoubtedly appeal to all those whose pursuits qualify them for membership. When it is remembered that practically all the benefits offered by the regular social clubs are provided in the Engineer's Club of Toronto, on a reasonably comfortable and elegant scale, and at a fee much less than is ordinarily required by such clubs, it is apparent that the limit of membership is likely to be reached very soon after the re-organized club opens its doors.

THE TEMISKAMING S.P.S. DINNER.

The rapid advance of mining interests in Northern Ontario, and the equal rapidity with which many of our graduates realized the value of the same, were spoken and unspoken lines of thought at a gathering held by them on December 20th., the event of the First Annual Temiskaming S.P.S. dinner. Although the number present did not include all graduates who resided in the district, the dinner was well attended, a huge success, and thoroughly enjoyed.

The toast to "The King" was proposed by Mr. Robert Bryce. In proposing the toast to "The Faculty and Engineering Professions," Mr. C. H. Fullerton drew attention to the increasing importance of the industry of mining, and expressed it as the unanimous opinion of graduates in the Temiskaming district that, in view of the fact that there is at present only one professor and no lecturers in the Department of Mining of the University, something should be done towards increasing the staff of this Department. He reinforced his remarks by stating that this year the class in fourth year Mining, numbers nineteen, and by comparing it with the same class three years ago, consisting as it did of two members. The toast was fittingly responded to by Mr. A. D. Campbell, president of the Engineering Society.

Mr. H. T. Routley proposed the toast to "Our Guests." In replying, Dean Galbraith congratulated the graduates on their achievements in the north country and thanked them for the reception accorded him, and the extreme pleasure he experienced as their guest. He related some interesting reminiscences of a trip into the Temiskaming country some six years ago, before mining in Cobalt had come into existence.

"Sister Institutions," proposed by Mr. E. V. Neelands, was responded to by Mr. W. S. Dobbs, representing Queen's Unievrsity. Mr. A. A. Cole, who was to have represented McGill, was unable to be present. Messrs. Thorne, Jupp, and Sutcliffe were called upon to respond to "Our Wives and Sweethearts" proposed by Mr. B. Neilly.

The dinner was such a splendid success that its annual recurrence will undoubtedly be looked forward to with interest.

In addition to the banquet, the visit of Dr. Galbraith comprised a trip underground at the O'Brien and Crown Reserve mines, and through the concentrating mills at the O'Brien, Silver Cliff and Nova Scotia properties.

It might be well to cite a few figures from an agricultural pamphlet, published a few months ago by the Ontario Legislative Assembly, regarding the rapid growth of the Temiskaming District.

"So recently as six years ago there were but 2,000 people in Temiskaming, while now there are between 50,000 and 60,000.

Englehart, a divisional point of the T. and N. O. Railway, is

25 miles north of New Liskeard; only four years old, it has a population of about 800, and is a promising agricultural centre. Westward, at the terminus of the branch line of the T. and N. O. from Englehart, is Charlton, a thriving village at the foot of Long Lake.

New Liskeard, on the T. and N. O., at the head of Lake Temiskaming, is 113 miles north of North Bay, and 340 miles north of Toronto; its population is 3,000, it has several important industries, and is in the midst of a large, well-settled agricultural district.

Haileybury, five miles to the south, is an attractive residential town, with a population of 4,000.

Cobalt, four or five miles farther south, has world-wide fame for its deposits of silver, being one of the most important mineral deposits discovered in the last forty years. Population, 5,500. Farther south is the growing town of Latchford."

It is also interesting to note that practically four-fifths of the land surveying in Temiskaming District is done by "School" men. The firm of Routley, Summers, and Malcolmson have surveyed over 700 mining claims during the past two years. Sutcliffe and Neelands, besides doing a large land survey practice, are municipal engineers of Cobalt, New Liskeard and Cochrane.

In another department is given a list of graduates in this section of Northern Ontario.

METHODS OF SECURING MAXIMUM EFFICIENCY IN
MANUFACTURING AND CONSTRUCTION.

By William M. Towle, B.S.*

Much is being said and written about conservation of natural resources. It is time that the American people paid more atten tion to the preservation of their heritage. With the great extent of territory and wealth of natural resources a tendency to run over and pick out the best and leave the rest to destruction and waste has been developed. In the manufacturing industries with the abundant supply of raw material and good home markets this same tendency is shown. Now when protection is lowered and the markets of the world are attracting attention it is necessary to consider the costs of manufacturing in all the details.

The cost of the manufactured product is made up of the cost of the raw material, labor, supervision, interest, and depreciation of plant, and expense of operating. In the ultimate analysis of the cost it will be found that labor represents nearly the whole amount, for the raw material before any labor is put upon it is but a small fraction of the whole cost of the article. Therefore, to reduce the cost of the finished product it is necessary to increase the efficiency of labor as no one wishes to reduce the wages of the laborer. Many schemes have been thought out and tried to accomplish this desired result. The most effort has been put on the wage side of the problem. As a general thing these schemes have been successful only when carried out by their originators or by some one in full sympathy with the plan.

The principal plans for rewarding labor are the day's-work, piece-work, premium or bonus, and profit sharing. The day's-work plan by which the laborer is paid a stipulated sum for the hour, day, or week, is the plan liked best by the workmen, for as a general thing they are willing to give a fair day's work for a fair day's pay. They look upon all other plans as methods to get more work out of them for the same money. While this is true to a certain extent, the other plans are mutually beneficial. The workman who is ambitious and willing to work harder gets more wages, and the manufacturer gets more work done with the same plant, thereby increasing the profits on the capital invested.

The profit-sharing plan, the most beautiful in theory is perhaps the poorest in practice. There are drawbacks in making a stockholder of an employee. While he is doing a good day's work the other workmen will criticise him as trying to "set a pace" and during a strike he loses influence with the other

* Professor of Industrial Engineering, Clarkson Memorial School of Technology.

workmen.   The average workman is ready enough to share in
the profits, but not to share in the risks of industrial or commercial enterprises.   The workmen have exaggerated ideas of
the profits. and not having access to the books are apt to think
that they are not getting their full share.

Some manufacturers share their profits by making an out-
and-out gift once a year to all of their employees who do not
in any other way share in the profits.   Generally these gifts are
from 5 to 10% of their annual earnings.   Another form of profit-
sharing is a pension, and a sick relief plan for the benefit of
the workmen and their families.   This is eminently proper, for
each industry should bear its own burden caused by accident,
sickness. disability, and old age.

Most employers do not know what the men and machines
are capable of doing.   If working on the day's-work plan the
men are liable to set an easy pace and the production be far
below the amount possible.   If a piece-work system is installed
the workman is left to devise ways of increasing the output.
When the earnings become large the price is cut and the usual
dissatisfaction and trouble ensue.

The greatest gain in securing the maximum efficiency in
manufacturing can be obtained by conserving the energy of
the workmen.   This can be done by investigating and studying all the operations performed by the workman, in designing
and arranging the tools. machines. and appliances; and in planning the method of doing the work that the workman may
produce the greatest amount of output with the least expenditure
of effort.   This can be secured only by the hearty co-operation
of the employer in providing the best of everything wherewith
to do the work, and by the workman in using such to the best
advantage.

Much has been done by several engineers who have made
a systematic and analytical study of each and every motion and
operation required to produce a given result. and who have
devised means and methods to produce the best and quickest
results at the least cost.   In Mr. F. W. Taylor's paper. "The
Art of Cutting Metals," before the American Society of Mechanical Engineers, the topics such as developing a high-speed steel,
determining the proper shape of the cutting tools. and deciding
about speeds, feeds, etc., were but incidentals to the study and
development of the larger problem.   Mr. Taylor's method involved a study of the various operations of a job and in timing
these operations; then in changing the conditions in accordance
with his time studies. until the minimum time in which the best
worker could perform them was determined; and finally in com-
pelling all the workers to conform to the methods of the most
skilful operator. and to equal his time. by means of bonuses
and penalties.

Every operation, for instance, is made up of a series of motions on the part of the worker. A careful study of these motions will eliminate all the useless movements of both man and material. This will give a set of standard operations to be followed thereafter.

It is thus seen to be a large problem to secure maximum efficiency in manufacturing. It involves providing the proper tools, appliances, and surroundings, and what may be considered the most important of all, namely, the employment of workmen of the proper physical and mental constitution to carry out to the letter the instruction given them for making the standard motions.

The average employer thinks that this is too large a problem to be studied seriously. He thinks it impracticable of execution. In this, however, he is very much mistaken. A series of articles by Mr. Frank B. Gilbreth, in "Industrial Engineering," shows how the adaptation of the standard motions to the trade of bricklaying has increased the efficiency of the men so that they can lay many more brick per hour and earn much more money per day, and to such an extent that for men working under the old methods there is hardly any comparison possible.

Mr. Taylor has worked out his system in some of the largest and most successful shops in the country. Mr. Gilbreth is likewise successful as a contractor, and his men all work according to the standard motions developed by him. The methods of both of these men have therefore withstood the test of commercial use.

Thus we see that to enable the manufacturer to turn out more and better products, at less cost, more attention must be paid to production engineering. New methods, tools, and facilities are expensive, and will be profitable only when the production is large. The overhead and distributing expenses of any manufacturing establishment are great; therefore, there must be a large output to share this burden in order to keep the cost per unit small. This has a tendency to work along the line of division of labor. The manufacturer becomes a specialist in a few things instead of producing a large line of goods.

It is this tendency towards specialization which has caused so many companies manufacturing similar products to unite their forces. Each factory confines itself to a few lines of work for which it is best fitted. This method eliminates ruinous competition in distribution, reduces the overhead and operating expenses, increases the efficiency of each plant, lessens the cost of production, and secures maximum efficiency in manufacturing and construction.

WHAT OUR GRADUATES ARE DOING

J. H. Caster, '07, is in the Production Dept., Canadian General Electric Co., Peterboro.

Walter Jackson, '07, is with the Ontario Power Co., Niagara Falls, as field engineer on construction and extension of works.

S. B. Iler, '08, is with the Seymour Power and Electric Co., Belleville, Ont.

A. D. Dahl, '08, is chemist with the Don Chemical Co., Midland, Michigan.

P. H. Buchan, '08, is in the Maintenance of Way Department, B. C. Electric Ry. Co., Ltd., Vancouver.

H. F. Shearer, '08, is testing engineer in the power apparatus Dept., Bullock Electric Mfg., Co., Norwood, Ohio.

N. G. Madge, '08, is chief chemist with the Continental Rubber Co., of New York.

F. C. Lewis, '08, is designing and estimating in the Railroad Bridges Dept., of the American Bridge Co., Chicago.

C. B. Langmuir, '09, is sales engineer for Factory Products, Limited, Toronto.

Stan. Stroud, '09, is with the Westinghouse Elect. and Mfg. Co., Pittsburg.

C. G. Titus, '10, is manager of the Bartlett Mines, Gowganda.

R. J. Spry, '10, is metallurgist for the B. C. Copper Co., Greenwood, B. C.

H. C. Bingham, '10, is assistant engineer in City Engineer's Dept., Moose Jaw, Sask.

L. J. Ireland, '07, is superintendent of Midland Construction Co., Ltd., in charge of the Seymour Power and Electric Co's construction work.

G. G. Bell, '05, is with Sawyer and Nuulton, of Portland, Me., as designer in power plants, pulp mills and general engineering projects.

C. J. Harper, '09, is in the employ of the General Electric Co., in Pittsfield, Mass.

E. A. Thompson, '09, is in the electrical department of the office of Smith, Kerry and Chace, Toronto.

L. H. Robinson, '04, is resident engineer for the Transcontinental Railway at Superior Junction, Ontario.

W. A. Begg, '05, has been connected with the Department of Public Works, Saskatchewan, on drainage surveys and road divisions.

E. E. Mullins, '03, has for some time past been mechanicel engineer for the United Fruit Co., operating in Central America and the West Indies. The company owns and operates some 2,000 miles of railroad (about 600 miles of which are in Costa Rica), and 90 steamships. Mr. Mullins has the supervision of repairs and improvements of the mechanical equipment of the railway and steamship lines.

˙APPLIED SCIENCE

INCORPORATED WITH

Transactions of the University of Toronto Engineering Society

DEVOTED TO THE INTERESTS OF ENGINEERING, ARCHITECTURE
AND APPLIED CHEMISTRY AT THE UNIVERSITY OF TORONTO.

Published monthly during the College year by the University of Toronto Engineering Society

EDITORIAL

Almost daily the press relates the death by accident of one after another, who has overreached his equilibrium in the atmos-

Aviation's Progress is Costly. phere and has become a victim through "the feel of the air." "Suddenly the machine struggled from his control, skidded, and cast itself downward to a heap of ruins."

Such passages meet our reading gaze so frequently, now-a-days that we wonder if the glory of achievement is really worth the sacrifice. Disaster seems to befall expert and beginner alike. In the majority of cases the cause will never be known. An imperfection in the mechanism, a fatal vagary of the upper air currents, or of the cast-iron will power of the operator, due to nervous tension, any or none of these may result in an added fatality to the already lengthy list.

Although the days of knight-errantry are a part of history,

reflected but seldom, one might easily form a mental comparison of the jousts and tournaments of old with to-day's passion to excell in the aviation world.

In the time of automobile infancy there were also fatalities, due to the same degrees of ignorance, impractibility and each of caution that summarize the present dangers of mechanical flight. But these were suppressed by strict enforcement of judicious rules and specifications—the result was a speed transition of the automobile from the sportsman's experiment to an important accessory to the business world. It is safe to say that its rapid advance was partly due to the timely laws which governed its manipulations.

There are few of us who do not see in the aeroplane, even a greater aid than in the automobile. Aerodynamics has long been a subject for deep study and experiment. Men have seriously attempted to build flying machines since the beginning of the sixteenth century, and altho the present revival of interest in acronautics has witnessed the accomplishment of a very great deal, mechanical flight may be held in abeyance, unless steps are taken to lessen the likelihood of loss of life and limb. In this regard Glenn H. Curtiss, one of the world's foremost aviators, writes:

"All efforts to improve and develop aviation will be futile. unless immediate steps are taken to guide and direct this development. The government, assisted by aero clubs, and aviators of experience, should take the matter in hand and guard against recklessness, which not only involves the lives of the aviators but also endangers the lives of spectators. Rules and regulations will not accomplish this purpose. The air is a great highway that has just been opened to the public. It must be regulated as are other great highways."

Although aeroplanes have been a marketable product for about a year, it is a question whether the industry a decade hence will in any way compare with the automobile business to-day unless the great degree of hazard is soon eliminated.

Among our graduates F. W. Baldwin. '06, and J. A. D. McCurdy, '07, have met with good measure of success in aeroplane manipulation. Both are members of the **S.P.S. men as** "American Aerial Experiment Association" **Aviators.** organized by Dr. Alexander Graham Bell, and of which Lieut. Selfridge, who met death in an aeroplane accident with Orville Wright. in 1909, was secretary.

Last Autumn, Mr. McCurdy arrested the attention of the scientific world when, while operating his machine at an aviation meet at Sheepshead Bay, he successfully transmitted the first wireless despatch from an aeroplane. His message read thus:

"Another chapter in aerial achievement is recorded in the sending of a wireless message from an aeroplane in flight."

Although his investigations have not resulted in any world beating aerial flights, Mr. C. H. Mitchell has devoted much thoughtful study to the science of flight, and has kept closely in touch with the progress that has been made to date. Mr. Mitchell will present a paper on this subject to the Engineering Society, at a general meeting in the near future. Undoubtedly the lecture will be one of the most interesting of the year, and those with whom the idea of mechanical flight is popular, are invited to be present.

CLASS '05 DINNER.

A Re-Union dinner of the Post-Graduate Class of 1905 and the Graduating Class of 1904 of the Faculty of Applied Science and Engineering was held on Wednesday, December 28th, 1910, at the St. Charles Hotel, Toronto. Twenty-six of the ninety original members of the two classes were present. Dean Galbraith was the guest of honor and in reply to the toast to "The University and the Old School" spoke reminiscently of the foundation of the School of Practical Science and of his connection with it in the early days. Mr. Kenneth Rose, '88, having heard of the dinner in Cobalt, dropped in unexpectedly and was accorded a hearty reception. Messrs. C. E. Bush and J. H. Craig of the "Science Octette" contributed music for the occasion. During the evening it was decided to hold another re-union in five years and the following list of new officers was elected to carry out this project: President, E. A. James; Vice-President, W. F. Wright; Secretary-Treasurer, W. W. Gray; Councillors, A. M. Campbell, A. Gray, R. S. Smart, B. B. Tucker. The chair was occupied by the retiring President, Mr. C. R. Young.

At a joint meeting of the Engineers' Society of Nova Scotia. and the Nova Scotia Institute of Science, in December. Mr. A. V. White, a graduate of '92, spoke to some length on the early history of Astronomy, including an investigation of the proofs usually given in support of the fundamental tenets of the copernican system. The discussion proved a valuable one to those who were fortunate in being present. and showed Mr. White's thorough familiarity with the history and theory of the subject. Mr. White has recently been making some engineering investigations in that province for the Conservation Commission of Canada.

The preparations for the twenty-second annual dinner of the Engineering Society are well under way and indicate another
The addition to the list of successes of similar functions of the Society.
Annual Last year the dinner was the most suc-
Dinner cessful ever held, and it is the aim of the present management to surpass it, as well as its many predecessors.

The twenty-first annual dinner was an important event, owing to the fact that approximately one hundred and fifty members of the Canadian Manufacturers' Association were in attendance as guests, and departed with a much enlightened idea of the prominence of this Faculty of the University and its importance to them.

The twentieth annual dinner of the Society was almost equally important. On this occasion the guest list comprised a large number of members of the Canadian Society of Civil Engineers. To them the excellent development of the Faculty of Engineering was strongly brought out, and they carried away a better impression, perhaps, than the Society had anticipated.

This year the Engineering Society is inviting members of the Board of Trade from every part of the province. Without doubt this is another step in the right direction, as this large and influential body should be well aware of the progress and possibilities of the University of Toronto; and to the number that will be in attendance, our views regarding the desire and willingness of the University to increase the much-needed intercourse between both bodies, will, in all probability, be presented.

The January "Applied Science" of last year contained an article pertaining to the efforts that had been made to establish
a course in Ceramics at the University of
Ceramics Again Toronto. That the Clay Products Manufacturers' Association have not diminished their eagerness to further the enterprise, is indicated by the fact that the question arose and was again lively discussed at their annual convention in this city some little time ago.

President McCredie, in his address, drew attention to the pressing need of the establishment of a chair of Ceramics in the University, as the manufacture of clay products required young men with a thorough training in this branch of science. Other members, especially Mr. S. J. Fox, M.P.P., and Mr. J. S. McCannell, emphasized the demand for such men in the ceramic industry.

The attitude taken by the Board of Governors several years ago, when the report prepared by Dr. Ellis, and others, who had investigated the progress in a number of universities in the United States, was not a favorable one in the interests of the Canadian Clay Products Association and its proposal. Its decision came rather as a surprise, as the agitation had met with the approval of the Faculty of Applied Science and of the University Senate.

In the article mentioned, Mr. D. O. McKinnon states clearly the scope such a course would have and what it might accomplish.

Those who are interested in the manufacture of clay, products are asked to again read the article and to look forward to a similar article from another member of the Association.

THESES

The fourth-year men, with the exception of the chemists and miners, have submitted their theses for examination. A list of them is given below:

E. P. Bowman—Magnetic Declination, Inclination and Intensity.

A. E. Glover—Water Supply and Irrigation with special attention to methods in practice in the United States.

J. S. Galletly—Photo-Topography.

J. E. Gray—Determination of Latitude.

O. W. Martyn—Determination of Longitude.

E. A. Neville—Theory and Practice of Road Location.

N. C. Stewart—Determinate of Latitude.

W. G. Amsden—The Construction of the Engineers' Transit and Level.

J. A. Baird—Drainage.

Thos. Barber—Dams and Forebays.

H. A. Barnett—Power Development.

W. M. Carlyle—Concrete Foundations, their Construction and Design.

H. S. Clark—Rock Excavation in Road Work, Tunnelling, etc.

C. G. Cline—River Discharge for Hydraulic Power Purposes.

G. A. Colquhoun—Street Paving.

R. L. Dobbin—Steel and Iron Bases for Columns.

A. W. Fletcher—Concrete Culverts.

W. C. Foulds—Iron Castings.

A. Fraser—Malleable Cast Iron.

M. M. Gibson—Dams, their construction and design.

V. A. E. Good—Fireproof Floors.

N. J. Harvie—Turbine Governors, their design, and methods of operation.

R. H. Johnston—The Purification of Sewage.

J. C. Keith—River Discharge.

G. A. Kingstone—Specifications for Concrete Reinforcing.

S. Knight—The Construction and Maintenance of Public Highways.

J. N. Leitch—Piles and Piling.

J. A. MacDonald—Reinforced Concrete Floor-Slabs.

J. B. Macdonald—Specifications for Structural Steel Work.

G. A. Macdonald—Irrigation—A description of the construction and application of various systems of design.

D. D. MacLeod—Reinforced Concrete Dams—their construction and design

S. G. McDougal—Permanent Railway Structure and Estimates.

J. McNiven—The Theory and Design of Compression Members.

W. H. Marten—Water Filtration, dealing with the most recent developments.

E. S. Martindale—Breakwaters—A treatise of their construction and efficiency.

F. S. Milligan—Bituminous Pavements—Construction of various types.

F. R. Mortimer—The Principles and Practice of Railway Signalling.

A. H. Munro—Preliminary Conditions in the Development of Hydro-electric Power.

V. A. Newhall—The Practice of Irrigation in the West.

F. T. Nichol—Reinforced Concrete Floor Construction—Various forms of construction and design.

C. M. O'Neil—Construction and Properties of Reinforced Concrete Columns, Beams, etc.

R. B. Pigott—Reinforced Concrete—Principles governing construction and design.

R. B. Potter—The Making of Good Roads.

W. S. Ramsay—Water Supply and Water Purification.

C. E. Richardson—Dams—General description of the construction of dams.

J. C. Street—The Purification of Public Water Supplies.

C. C. Sutherland—Sewage Disposal—A description of methods now in common use.

A. D. Sword—Diving—together with other applications of compressed air to submarines.

L. T. Venney—Concrete Highway Bridges.

C. M. Walker—Cement—its properties and the various methods employed in its manufacture.

G. A. Warrington—Highways—with special attention to location and surfacing.

L. A. Wright—Concrete Bridges.

W. S. Young—Road Improvement.

D. C. Blizard—Tungsten Lights, with reference to Feature, Field, and Effect.

L. S. Cockburn—Storage Batteries—The application, installation, care and maintenance of batteries; synopsis of the various systems in use.

A. G. Code—The Development of the Incandescent Light—from the age of experiment to its present efficiency.

C D. Dean—Suction Gas Producers—their development and construction; general description of plants and units comprising them.

W. P. Dobson—Wave Forms of Force Periods which occur in Alternating Current Circuits.

J. W. Ferguson—Suction Gas Producers.

F. T. Fletcher—The Development of the Incandescent Lamp, with reference to the Tungsten Lamp.

C. B. Leaver—Suction Gas Producers, Gas Analysis, etc.

H. G. MacMurchy—Modern Methods of Electric Power Distribution.

H. J. MacTavish—Integrating Watt-meters—Principles, construction, and application.

C. Holmes—Storage Batteries.

D. D. McAlpine—The Induction Generation, its theory, application and principles of construction.

H. O. Merriman—The Protection of Electric Circuits from Lightning.

P. E. Mills—The Gasoline Engine as applied to Modern Automobiles.

C. G. Parker—High Voltage Transmission.

A. L. Sutherland—The Prevention of Smoke in Boiler Plants.

R. M. A. Thompson—The Electrification of Steam Railways, together with a description of the various types of railway control.

K. M. VanKallen—Oil Switches.

M. B. Watson—The Electrical Equipment of a Modern High Tension Distribution Station.

B. W. Waugh—Wireless Telegraphy.

F. C. White—Voltage Regulation—the application of vector diagrams to the same.

G. K. Williams—Gasoline Carburettors.

A. W. Youell—Air Compressors.

S. E. Craig—Pulp Making Machinery.

J. R. Burgess—Gas Producers—A description of various types in use, their operation and efficiency.

A. W. Chestnut—Refrigeration—Application of the general principles to refrigeration requirements, and a description of the machinery in use.

D. C. Chisholm—The Preservation of Timber.

C. R. Ferguson—Steam Turbines—The theory governing their application, the field, and efficiency.

H. Gall—Municipal Railways and Thoroughfares.

V. F. Gourlay—Tool Steel—Its properties, manufacture and a synopsis of the various uses to which it is applied.

H. C. Johnston—Dams—Design and various forms of construction.

J. I. MsSloy—Modern Automobile Engines.

C. H. Phillips—Producer Gas Engines.

L. R. Wilson—The Application of Nickel Steel and other High Tension Alloys to Modern Bridge Construction.

M. H. Woods—The Details of Gasoline Engine Details.

H. C. Barber—High Tension Underground Cables.

T. H. Crosby—The Parallel Operation of Alternating Current Generators.

R. G. Lee—The Single Phase Induction Motor.

A. S. McCordick—Railway Converter Sub-Stations.

C. E. Palmer—Storage Batteries and their Application.

K. K. Pearce—Electric Locomotives—A description of various types of electric train control, and of modern brake systems.

R. A. Sara—Organization for Electric Distribution Companies.

A. Schlarbaum—Protective Apparatus.

M. W. Sparling—Switchboards—A synopsis of their construction and equipment.

J. B. K. Fisken—A Gothic Church.

T. C. McBride—The Design of a Public Library.

W. S. Wickens—The Design of a Post Office.

C. P. Van Norman—Transformers—A description of general principles, and the investigation of vector relations of the transformer.

R. J. Arens—The Manufacture of Potassium Bichromate from Chrome Ores.

J. H. Harris—The Manufacture of Soap.

J. A. MacKinnon—The Filtration of Water for Domestic Purposes.

J. H. Adams—The Limits of Amalgamation of Porcupine Gold Ores.

E. T. Austin—High Silica Slags in Copper Smelting.

D. G. Bissett—The Electrolytic Assay of Zinc Ores.

A. F. Brock—The Cyaniding of Concentrates of Porcupine Ore.

A. D. Campbell—Cobalt Milling Methods.

V. H. Emery—Cyanide Treatment of Slimes from the Porcupine Gold Ores.

J. M. Foreman—Acid Open Hearth Slags.

R. L. Greene—Association of the Ores at the O'Brien Mine Cobalt.

P. E. Hopkins—Cyanide Treatment of Sands from Porcupine Gold Ores.

F. L. James—The Electro-analysis of Zinc.

H. G. Kennedy—Mine Timbering of Metalliferous Mines.

G. L. Kirwan—Losses in the Fire Assay of Silver Ores.

A. W. R. Maisonville—Malleable Castings.

A. C. Matthews—The Basic Open Hearth Slags.

W. E. Newton—The Electro-magnetic Separation of Zinc Minerals.

A. L. Steeel—Investigation of the Losses in Assaying of Gold-bearing Ores.

H. M. Steven—Cyaniding versus Amalgamation for Gold Ores.

T. Walton—Blast Furnace Slags.

S. A. Wookey—The Interpretation of Topographic Maps, with some reference to the Geology.

SCHOOL MEN IN THE NORTH

Some of the "School" men at present residing in the Temiskaming District:

In Cobalt—Chas. Williams, M. T. Culbert, Robt. A. Bryce, A. Carroll, B. Neilly, E. V. Neelands, S. A. Thorne, H. E. Colley, Herb. Clark, G. Pace, D. Duthie, A. D. MacDonald, Geo. Adams.

In New Liskeard—E. W. Neelands, H. W. Sutcliffe, C. H. Fullerton, F. Dines, P. Phillips, W. J. Johnston, W. J. Blair.

In Haileybury—H. T. Routley, W .Malcomson, G. F. Summers, Jno. Pierce, E. D. Eagleson, A. H. A. Robinson, A. E. Jupp, Mr. Gouldie, H. Southworth, Eric Ryerson, Robt. Laird, Geo. Johnston.

In Cochrane—R. Harthstone, E. Nickels, P. A. Laing, W. M. Bishop, J. H. Dawson, Gordon Calvert. Chas. Peterson, Erie Ryerson, Buzz Nelles.

In Porcupine—E. R. Dann, F. Bedford.

In GowGanda—C. Titus, Wilkie Evans.

In North Bay—R. Keys. P. Maher. Harold Keefer, Jno. Shaw.

In Sault Ste. Marie—Jno. Laing. Ken. Ross.

In Sudbury—Mr. Stull, Mr. Demorset.

OBITUARY

We have to announce the death in Buenos Ayres, Argentine Republic, of Roy B. Ross, '05. Mr. Ross was 26 years of age, and had been in Buenos Ayres for two years in the interests of the Marine Signal Co., Ottawa. A full obituary will appear in the April issue.

WHAT OUR GRADUATES ARE DOING.

F. G. Allen, '07, is with the B. F. Sturtevant Co., Hyde Park, Mass., as assistant to the chief consulting engineer.

A. H. Arens, '06, is in Inverness, N.S., as resident engineer for the Inverness Railway and Coal Co.

E. G. Arens, '09, is with the H. D. Symmes Co., of Niagara Falls.

A. R. Raymer, '84, is assistant chief and signal engineer for the Pittsburgh and Lake Erie R. R. Co.

Wm. J. Chalmers, '89, is with the Vanport Beaver Co., Pittsburg.

G. H. Richardson, '88, is managing director of the Yellowhead Pass Coal and Coke Co.

G. E. Sylvester, '91, is chief engineer for the Canadian Copper Co., Copper Cliff, Ont.

A. T. Beauregard, '94, is laboratory engineer for the Public Service Corporation of New Jersey.

Harold Rolph, '94, is secretary for the John S. Metcalf Co., Limited, Montreal.

A. E. Blackwood, '95, is manager of the Sullivan Machinery Co., New York city, manufacturers of quarry and mining supplies and diamond drills.

H. W. Saunders, '01, is division engineer of the United States Coal and Coke Co., having charge of 12 of their mines.

Wm. Christie, '02, is engaged in D.L.S. work at Prince Albert, Sask.

A. A. Wanless, '02, is lecturer and demonstrator in the Sydney Mines Technical Schools, Nova Scotia.

A. G. Lang, '03, is underground superintendent for the Toronto Hydro-Electric System.

H. F. White, '03, is assistant superintendent for the Geo. White & Sons Co., Ltd., London, Ont.

John Paris, '04, is resident engineer on construction for the T. C. Ry., at La Tuque, Quebec.

W. S. Pardoe, '04, is instructor in hydraulics and sanitary engineering, University of Pennsylvania, Philadelphia.

Thos. D. Brown, '06, is with the Canadian Fairbanks Co., in their Calgary office.

W. P. Near, '06, is resident engineer on trunk sewer construction, Main Drainage Dept., City Hall.

Geo. W. Bissett, '06, is mill superintendent for the Canadian Exploration Co., Ltd., at Naughton, Ontario.

D. W. Marrs, '06, is designer of structural steel for the Riter Conley Mfg., Co.

M. K. McQuarrie, '07, is terminal engineer for the C.P.R. at Vancouver, B.C.

Drawn by T. L. F. Rowe when a First Year Student in
the Department of Architecture.

Applied Science

INCORPORATED WITH

TRANSACTIONS OF THE UNIVERSITY OF TORONTO ENGINEERING SOCIETY

| Old Series Vol. 23 | FEBRUARY, 1911 | New Series Vol. IV. No. 4 |

INCANDESCENT LAMPS.*

M. B. HASTINGS. '11.

Illuminating engineering is simply the engineering method applied to the production, the sale and the use of light. Now, as this paper must be short I have decided to confine my few remarks to the manufacturing and testing of lamps. Hence the title " Incandescent Lamps."

Incandescent lamps, as we are all aware, are divided into four separate and distinct classes, namely, Carbon, with an efficiency of 3.1 w.p.c.; Gem, which is an improved carbon with an efficiency of 2.5 w.p.c.; Tantalum with an efficiency of 2 w.p.c.; and Tungsten with an efficiency of 1.25 w.p.c.

As the Carbon lamp is the first to come into prominence it might be proper to mention a few words with regard to its manufacture. The filaments of the first carbon lamps consisted of strips of woody plants such as bamboo, and vegetable fibres, such as cotton threads, and strips of paper and animal fibres, such as silk, but none of these materials mentioned were found sufficiently uniform in structure, density, electrical resistance and character of surface. The first promising effort to gain uniformity was the effort to parchmentize the outer portion of the cotton threads usually by soaking a limited time in sulphuric acid, but this resulted in a fibre covered with lumps and irregular in section. This objection was overcome by drawing the filament through a series of dies. This process pointed the way to the evidently logical procedure of breaking up the cellular structure entirely, converting it into a viscous compound which could be pressed through an orifice into a coagulent, thus forming a thread of cellulous without structure. Any chemical capable of dissolving cotton may be used, a strong hot solution of zinc chloride being the most common. A solution of gun cotton on acetic acid is used, to some extent, principally in European factories.

The process of dissolving the cotton for carbon filaments in a zinc chloride solution takes the form of a heavy syrup. It is then evacuated at a temperature below the boiling point of water

---

*Read before the Engineering Society, January 25th, 1911.

until occluded air and gas are removed. All conditions must be uniformly maintained. After cooling the mass is forced through platinum dies by air pressure, the end of the dies extending about ½ in. into a jar of alcohol, which coagulates the jet of solutions as it emerges from the die. The glass jar containing the alcohol revolves about one revolution in 3 min. which keeps the fibre from tangling as it lays itself down in coils. Inside the glass jar is a hard rubber basket into which the fibre is coiled. One of these baskets will hold enough fibre for 1800 — 2 c.p. 110 volt lamps while it will only hold enough for about 300 — 50 c.p. 50 volt lamps. After standing in alcohol for about 12 hours these fibres are washed, first in a solution of hydrochloric acid water for about 3 hours, then in distilled water for about 2 hours. The idea is to dissolve out all soluble salts from the fibre, leaving only pure carbon with traces of insoluble salts. The latter are afterwards evaporated by the heat of carbonization. If the salts were not almost wholly removed from the fibre the carbon would be porous and brittle. After washing the fibre is wound upon drums about 4 ft. in diameter where it dries. After the fibre is taken from the drum it is wrapped either upon carbon blocks or forms to give the desired shape, and carbonized at a temperature of about 3000 degrees Centigrade. After carbonization the filaments are assorted, cut for length and electrically heated in the presence of paraffine hydro carbon gas. The carbon deposited upon the filament from this gives it a better light radiating surface, also the deposit causes the resistance of the individual carbons to be brought closer together. The amount of deposits is governed exactly by automatic devices. After the carbons have received this coating they are put in stock for factory use. The glass tubing from which the stems containing the platinum leading in wires are made is bought from the glass factory in 3 or 4 foot lengths. This tubing is cut into lengths varying from 1 to 2 in. on a carborundum wheel. These pieces are then placed in an automatic flange rolling machine. From these machines the flanges are taken, assorted and inspected. They are then taken by the stem machine operators, and with the platinum and copper leading in wires together with the anchor wire they are made into stems which after inspection are ready for the clamping department. In this department the stem and filament meet. The operators cement the carbon to the platinum wires by a graphite cement, after which they are baked until the joint will stand soaking in water for 4 or 5 hours without coming loose. We now have the carbon cemented to the platinum leading in wires, ready to be sealed into the bulb. They are then sealed into the bulb and the vacuum is obtained, similar to the process which will be described later in the discussion of the Tungsten lamp.

Tungsten Lamps.

Within the last few years wires of rare metals have been substituted for the older carbon wires or filaments. Tantalum and Tungsten metals have proven best adaptable for this purpose, and it is the Tungsten lamp especially that this paper is designed to discuss.

Tungsten lamps are manufactured from the ore known as Wolframite, an oxide of manganese, iron and Tungsten (wolfram). Tungsten oxide is now chemically obtained from the crushed ore. Metallic Tungsten, as a powder, is now reduced from the yellow oxide. The paste process by which a large majority of Tungsten lamps is made is as follows: Tungstic acid, as received by the lamp manufacturer, is in the form of a heavy yellow colored powder, and notwithstanding its purity has to undergo a most thorough purification ,and special treatment in order to reduce it to a fine yellowish powder which possesses a slight porosity. This peculiar physical condition is necessary for the successful reduction of the oxide which can be accomplished in several different ways, perhaps most easily by heating to redness in a current of hydrogen. After obtaining this very pure metallic Tungsten the next operation is to mix it with combining material in order to form a plastic mass that may be squirted into the fine threadlike filaments. Some substances can be used for this purpose, but some compound of carbon, oxygen and hydrogen, such as starch, sugar, camphor, etc., is usually employed. The metallic powder when mixed with such combining material has the same consistency and appearance as black putty. It is absolutely smooth and uniform and it is impossible to detect the slightest grain. This paste, as it is called, is placed in a small steel cylinder and forced by a pressure of about 32,000 lbs. per sq. in. through a small diamond die.

The die used in squirting Tungsten filament consists of a suitably mounted diamond of from one-half to one carat in weight, through which a very minute hole has been drilled. In the smaller dies used to-day this hole is only about 0.0014 in. in diameter, which is smaller than an ordinary hair. The hole is drilled in the diamond with a steel needle, ground down so fine that it is as flexible as a hair, and, as can be imagined, the drilling requires considerable time and patience. The stone when drilled is mounted in a steel casting in order to hold it against the enormous pressure used in squirting the filament.

Under such pressure the wearing of the die even by smooth tungsten paste is very rapid. This wearing is a serious matter, as the diameter of the hole, and consequently that of the filament squirted, constantly increases. Moreover, the wearing is not uniform, so that the hole enlarges more rapidly in the direction of one diameter than the other, assuming when worn an elliptical shape. After enough filament for about 1,500 lamps has been squirted it is necessary to have the die rebored, an operation which costs almost as much as the original die. A die cannot be rebored more than twice before it develops cracks or fissures which cause it to break. The next hardest material, sapphire, has been experimented with as a material for these dies, but it is found that such a die is very liable to split, and that it will hardly make 100 lamps before it needs re-drilling.

The filament, after squirting has been likened in character to a filament of putty, that is, while holding its form well and being flexible to some extent, it is liable to break if bent sharply. The filament as squirted is looped back and forth on cards, and after being allowed to dry is cut to form a number of single loops, much as they are seen in the finished lamp.

The next operation is to beat the filaments in an inert gas, or in one chosen to act upon the particular binding material employed, until they reach a red heat which removes any moisture and lighter hydrocarbons that may be present. Each filament is then mounted in a current conducting clips, so that it may be heated by passing an electric current through it. During this final heating or forming, as it is termed, the filament is supported vertically with the loop downward. A very small weight (a few milligrams) is hung in the loop to prevent the filament from being distorted in shape during the heating, which is usually performed in either inert gas or in a very good vacuum, the gas, if used, being again dependent to some extent on the binding material employed. The temperature of the filament is raised gradually, allowing time for the proper reactions and physical changes to take place. During the heating the energy put into the filament rises to about fifteen times that finally required in the lamp, and while some heat is carried away by the forming gas the temperature is undoubtedly much higher than that reached in subsequent operation.

Every trace of binding material is driven out, and the filament when finally brought to a sufficiently high temperature undergoes a sudden and marked contraction in diameter and length as the small semi-molten particles become soft enough to merge into one perfectly homogeneous mass. A piece of such filament under a microscope resembles a drawn wire, and while the surface is not perfectly smooth there is no indication of a granular structure.

At a dull red heat any good tungsten filament is flexible enough to be bent as desired, but when cold is somewhat fragile. For this reason it is a good thing, if possible, to light tungsten lamps while cleaning them, the chance of mechanical breakages being then minimized.

In order to secure a uniform quality of filament, which is absolutely necessary for good lamp making, every step of the entire process of production, even to the smallest detail, must be carried through with exactness. For example, the rate at which the temperature of the filament is raised during the forming process has a most important effect on the final filament structure, and must be carried out with extreme care in order to assure a perfectly uniform run of filament.

After forming the filaments are mounted upon the familiar glass supporting rod as seen in the finished lamp, and the joint between each leg and the supporting and conducting terminal made by electrically welding the filament to the support. This makes a very perfect joint, both electrically and mechanically.

The material used for the hooks and supporting wires affects the performance of the lamp to a considerable extent. Soft copper is used extensively for such supports, also to a somewhat lesser extent, molybdenum, tungsten, tantalum, carbon, platinum or other refractory materials. Soft copper does not alloy with pure tungsten, and moreover occludes but very little gas, so that it makes a very satisfactory form of supporting hook.

The process of producing a high vacuum within the bulb of a tungsten lamp is a great deal more laborious than in the case of the carbon lamp. It is necessary in order to produce a good lamp, and this applies to carbon and other types as well, to remove every trace of gas, not only that actually free at the time of pumping, but that which may later be liberated from anything within the finished bulb. A good lamp vacuum is only possible through the use of very perfect exhausting machinery and through subjecting the entire lamp, filament, glass, and other parts to a proper heat treatment, during the pumping process. The heat treatment tends to drive from the glass walls and other surfaces exposed within the bulb the particles of air which cling to them in a thin film with surprising persistence. A glass bulb pumped while cold will apparently reach a high degree of exhaustion at the end of the pumping process but, if left standing for some time, will be found to possess a very poor vacuum as judged from that required to insure a good lamp performance. This is due to the gradual liberation of particles of air which, during the pumping, cling to the interior surfaces. The filament is heated intensely by passing an electric current through it while on the pump in order to drive from it and from the supporting wires, which are heated by the incandescences of the filament, any occluded gas which may later be freed from these parts and thus spoil to a certain extent the perfect vacuum required for successful operation. During the pumping process the filament temperature must be regulated with great care, the temperature being raised gradually as the bulb becomes evacuated. If, for example, the temperature should be raised too quickly, a thin film of oxide will form on the surface of the filament, which although entirely removed by a further rise in temperature and higher degree of exhaustion, will have been found to have caused a slight change in the character of the surface of the filament, altering thereby its emissivity and radiating properties and injuring its subsequent life performance.

After pumping, the lamps are given an exhaust inspection and aging by burning at a certain percentage over voltage. If, during this inspection, any lamp develops or shows a bluish color or haziness within the bulb it is an indication of imperfect vacuum and the lamp is rejected. The period of burning and per cent. over voltage depends somewhat on the size of the lamp, but in every case has been chosen so that if a lamp is at all likely to develop a poor vacuum it will be disclosed on this inspection.

The glass work required in making the tungsten lamps, is, with the exception of the centre glass stem, practically the same as in the other types of incandescent lamps.

I have now covered in a general way the process of producing the tungsten lamps, and have, I hope, given some slight idea as to the care required in such production. In order to show the performance of the lamps in subsequent service I have at my disposal, loaned by the N. E. L. A., the averaged candle power

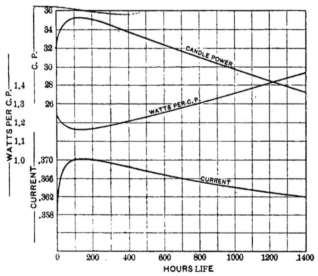

Fig 1 —Characteristic performance of 40-watt tungsten lamps

life curves of 50-40 watt tungsten lamps which were burned on life test at constant voltage corresponding to an initial consumption of 1.25 watts per mean horizontal candle power. The curves in Fig. 1 show the change in candle power, current and efficiency during the period of test, which was stopped at about 1,400 hours. In order to show the life performance, a curve is given showing the per cent. of lamps which were burning at the end of various periods of time. These curves, Fig. 2, are the average obtained from 80-40 watt tungsten lamps burned under same conditions as the previous test, i.e., at 1.25 w.p.c. About one-half of the lamps were burned in a horizontal position and others in a vertical position, tip downward. These tests were stopped at the end of 2,000 hours.

The change in appearance of the filament during life is

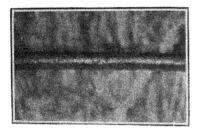

FIG. 3.—Unmounted tungsten filament.

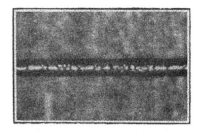

FIG. 4.—25-watt tungsten—400 hr.—alternating current.

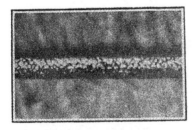

FIG. 5.—40-watt tungsten—180 hr.—alternating current.

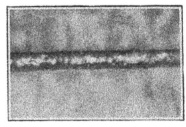

FIG. 6.—25-watt tungsten—1400 hr.—alternating current.

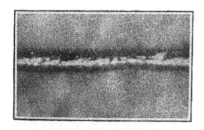

FIG. 7.—40-watt tungsten—898 hr.—alternating current.

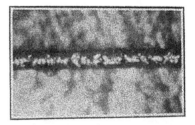

FIG. 8.—40-watt tungsten 2190 hr.—alternating current.

rather interesting, and some illustrations showing filaments taken from lamps which had been burned for various lengths of time are shown in Figs. 3 to 23. For comparison, a few tantalum filaments, as well as a few filaments from a gem and from a carbon lamp are shown.

To show the effect of a varying line voltage an increase of 6 per cent. in the voltage of a carbon lamp increases the per

cent. candle power as much as an increase of 9 per cent. in
voltage of the tungsten lamp. While a tungsten lamp of 10 per
cent. under voltage drops in per cent. candle power only as much
as a carbon lamp, 7 per cent. below normal. For a circuit where
the regulation is poor, the tungsten lamp will be found to give
much better satisfaction as far as variation of candle power is
concerned than the carbon lamp. The size and filament required

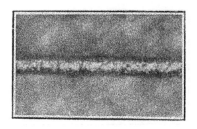

FIG. 9.—40-watt tungsten—300
hr.—direct current.

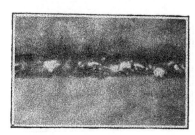

FIG. 10.—40-watt tungsten—3000
hr.—direct current.

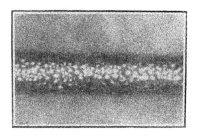

FIG. 11.—100-watt tungsten—
1400 hr.—alternating current.

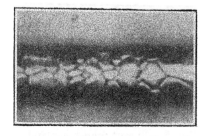

FIG. 12.—250-watt tungsten—
200 hr.—alternating current.

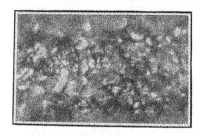

FIG. 13.—Tungsten—street series
—2000 hr.

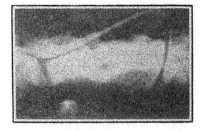

FIG. 14.—Tungsten—street series
—2349 hr.

in lamps of various candle powers and voltages varies considerably. For 110 volts the size and length varies from a diameter
of 0.0013 in. and a total length of 17.4 in· for a 25-watt lamp to a
diameter of 0.0060 in., and a total length of 37.56 in. for a 250-
watt lamp.

The specific resistance of tungsten is about 46.5 ohms per

mil-inch at a temperature corresponding to 1.25 w.p.c., and about 9 per cent. of this figure or 4.19 ohms per mil-inch at ordinary room temperature. This low cold resistance has led to a good deal of discussion as to the value of the starting current obtained upon closing the circuit of a tungsten lamp and its effect upon the filament.

The question is of particular importance when the lamps

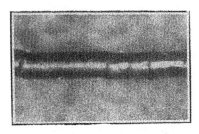

FIG. 15.—80-watt tantalum—800 hr.—alternating current.

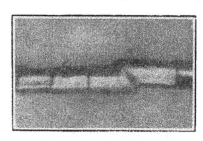

FIG. 16.—80-watt alternating—400 hr.—80-watt alternating current.

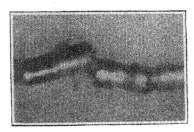

FIG. 17.—40-watt tantalum—800 hr.—alternating current.

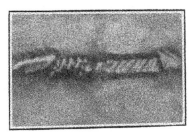

FIG. 18.—80-watt tantalum—948 hr.—alternating current.

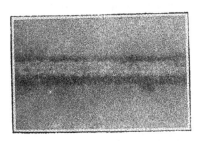

FIG. 19.—80-watt tantalum—60 hr.—direct current.

FIG. 20.—80-watt tantalum—455 hr.—direct current.

are turned on and off continuously, as for example, in flashing sign work.

To most people incandescent lamps are pear-shaped bottles of two or three sizes used for store and residence illumination. As a matter of fact, there are 435 sizes, styles and types of incandescent lamps, and they are used for every conceivable ser-

vice from street lighting to the illumination of stomach in-
teriors in surgical work. These lamps range from 1-10 to 400
c.p. There are 97 separate sizes and types of bulbs used and 26
sizes and styles of base. There are 22 different types of filaments. If all the different voltages, efficiencies, styles and types

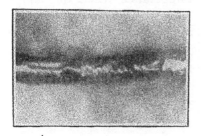

FIG. 21.—80-watt tantalum—
1058 hr.—direct current.

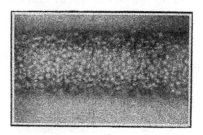

FIG. 22.—Gem—841. hr.

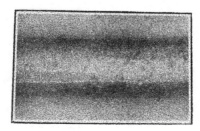

FIG. 23.—Carbon—508 hr.

are taken into account there are in practice not less than 2,500
distinct lamps.

It may be interesting to explain how characteristics of these
lamps are obtained, that is, the data necessary for testing these
lamps from a mathematical and scientific standpoint-basis.

The tungsten lamp is measured at a voltage which will make
it burn at 1.25 w.p.c., the candle-power, watts, amperes, ohms
and volts are plotted and as these readings are from the lamp
at its normal condition they are recorded as 100 per cent.

Corresponding readings are taken as the candle-power is
lowered to 20 per cent. normal and raised to 150 per cent. of
normal. For the normal range of operation these curves can be
expressed in the form of parabolic equations, a few of which are
given as follows.

Let c = candle-power.

e = efficiency in watts per candle.

V = voltage.
W = watts.

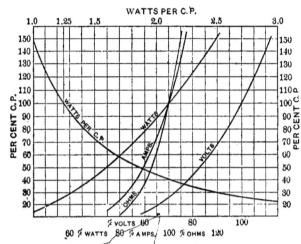

FIG. 24 —Performance of tungsten filament lamp

$$\frac{c_1}{c_2} = \left(\frac{e_2}{e_1}\right)^a = \left(\frac{V_1}{V_2}\right)^b$$

$$\frac{W_1}{W_2} = \left(\frac{V_1}{V_2}\right)^d$$

Values of exponents "a," "b," and "d" are obtained by plotting the corresponding curve on log section paper and measuring the tangent of the angle, which is the value of the exponent required.

These performance curves are of indispensable value in the engineering department in connection with life testing of the lamps sent in by the different factories.

. Within the last year the basis of candle-power has changed from the old American candle-power to the International candle-power. The effect of this change from the British unit of 1.6 per cent. in the unit of the Bureau of Standards, which is in general use for electric lighting throughout the country is to raise the candle-power rating and increase slightly the w.p.c. of electric lamps. A 16-candle-power lamp will give 16.26 candles in the new unit, or a 16-candle-power carbon filament lamp burning at 110 volts will give 16 candles on the new basis at 109.69 volts. The change though small is important in the photometry and rating of lamps.

Another recent change is what is known as the 3-voltage plan. The incandescent lamp manufacturers have recently made a radical change in their method of rating these "Mazda" lamps, in order that the lamps could be used with greater economy under those certain conditions where heretofore their cost of operating exceeded that of a less efficient type of lamp and is most valuable in cases where the cost of electrical energy is low. The new method of rating, called "The Three-Voltage Plan," is based upon the fact that for any given set of conditions, depending upon the cost of energy and cost of lamp there is one particular efficiency and life at which it is most economical to operate a given lamp.

Each "Mazda" lamp is labelled with three voltages, two

$$114$$

volts apart, as, for example, 112, called top, middle and bottom

$$110$$

voltages. This method of rating makes it possible for a customer to select the particular efficiency of lamp he wishes to use by specifying that either the top, middle or bottom voltage, as the case may be, should be the same as that of his lighting circuits.

When burned at top voltage the "Mazda" lamp has the highest efficiency or consumes the least energy for the light produced, and gives life of 1000 hrs. At middle voltage more energy is consumed per candle-power produced and the life is lengthened (due to operation at a lower temperature) to 1300 hrs. At bottom voltage the lamp is operating at the lowest efficiency and gives a life of 1700 hrs. It is obvious that the relative cost of lamp and energy will determine the most economical life and efficiency, since if energy is cheap the saving in energy obtained by operating the lamp at high efficiency is not sufficient to counter-balance the higher resulting renewal expense. On the other hand, if the energy is relatively expensive then it will be desirable to operate the lamp at a high efficiency, since the saving in current at the higher rate will more than pay for the increase in renewal expense.

The efficiency of the different sizes of lamps at top voltage is not the same, since the larger lamps are relatively longer lived than the smaller ones, and, in order to give all sizes a uniform life of 1000 hrs. at top voltage it was necessary to operate the 25-watt. lamp at 1.33 w.p.c., the 40-watt at 1.25 w.p.c., the 60-watt at 100 and 150-watt at 1.20 w.p.c., and the 250-watt and 500-watt at 1.15 w.p.c.

This three-voltage plan was adopted last May by the co-operative companies of the National Electric Lamp Association. Some time prior to this the word "Mazda" was agreed to by the companies forming the National Electric Lamp Association along with others as a word which will accompany the highest efficiency lamp that is made. At the present time the "Mazda" lamp is composed of tungsten filaments or some at present unknown filament which would result in a lamp of higher efficiency than the present tungsten filament lamp, it would be the "Mazda." "Mazda," therefore, is the hall-mark of quality and progress in high efficiency metal filament lamp manufacture, or the seal of the world-wide co-operative effort to produce the best.

A great many people seem to think that illuminating engineering and the different phases of work in connection therewith is narrow, but the "Mazda" lamp and its importance from an invention standpoint shows the broad field there is for the illuminating engineer.

To show its real value, let us consider a steam turbine, electric system and note the great loss in conversion of energy from coal to light.

From Fig. 25, we start with coal containing 100 per cent. energy. Consider boilers working at an efficiency of 70 per cent.

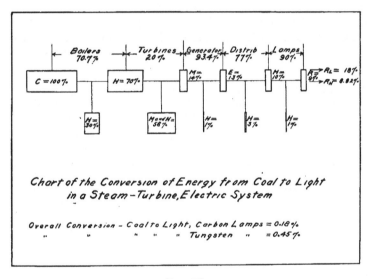

FIG. 25.

we have 70 per cent. going as useful energy to the turbines and 30 per cent. wasted as heat loss. Turbines at an efficiency of 20 per cent., we have 14 per cent. going to generations and 56 per cent. wasted in overcoming mechanical resistance and heat loss. Generations at the efficiency of 93 per cent., we have 13 per cent. going to the distributing system and 3 per cent. wasted. Distributing system 77 per cent. efficient, we have 70 per cent. going to the lamps and 1 per cent. wasted. Lamps 90 per cent. efficient, we have 9 per cent. of the 100 per cent. started with, as radiant heat and light therefore we have only 18 per cent. of what we started with a light when carbon lamps are used. Now let us substitute "Mazda" lamps, and, considering their relative efficiencies as carbon, 3.1 w.p.c. and "Mazda" 1.25 w.p.c., we would have 45 per cent. instead of 18 per cent. This remarkable triumph in the science of light can be further shown by imagining an almost impossible accomplishment, that is the invention of a turbine working at an efficiency of 50 per cent. Now, if such were the case, 3 per cent. of the original energy would go to the generators instead of 14 per cent. and considering the rest of the system the same as before there would be 32.5 per cent. available for the distributing plant and 25 per cent. would be radiant heat and light; we have 45 per cent. of the original 100 per cent. available for light when carbon lamps are used. Among mechanical engineers it is hardly expected that the existing efficiencies in machines will improve to a very great extent, therefore, the "Mazda" lamp is equivalent to an imaginary turbine of 50 per cent. efficiency and consequently opens a new field in the realms of engineering, and when we look back and find the carbon lamp has been improved from 6.5 w.p.c. to 2.5 w.p.c. in the Gem (which is an improved carbon filament), it is hard for one with a most vivid imagination to forecast its future when it starts at excellent efficiency of 1.25 w.p.c.

The following table shows comparative costs for obtaining 80 candle power from Carbon, Gem, Tantalum and "Mazda" lamps. At the present electric power rates, a large saving is noted in favor of "Mazda" lamps. At a power rate of 10c per kilowatt hour the cost for 80 candle power of "Mazda" light for 1000 hours burning (which includes the lamp cost) is about three sevenths of the actual cost of power alone for equal candle power from carbon lamps of 3.5 w. p. c.

Designation	Carbon		Gem	Tantalum	Mazda
Watts per Candle—Actual....	3.5	3.1	2.5	2.0	1.25
Size of Lamps............	16 c. p.	16 c.p.	40 c. p.	40 c. p.	80 c. p.
No. of Lamps Compared.....	5	5	2	2	1
Total Candle Power.........	80	80	80	80	80
Watts per Lamp............	56.0	49.6	100	80 ·	100.0
Total Watts for 80 c. p......	280	248	200	160	100
Hours Useful Life—Each.....	830	450	460	800	800
Cost of Lamps..............	$ 1.00	$ 1.00	$ 0.70	$ 1.70	$ 1.75
Cost of Renewals, 1000 Hours	1.205	2.22	1.474	2.126	2.188

Cost of Power per K.W.H.		Table of combined cost of Power and Lamp Renewals for 80 c. p. for 1000 Hours				
	.01		4.005	4.70	3.474	3.726	3.188
	.02		6.805	7.18	5.474	5.326	3.188
	.03		9.605	9.66	7.474	6.926	5.188
	.04		12.405	12.14	9.474	8.526	6.188
	.05		15.205	14.62	11.474	10.126	7.188
	.06		18.005	17.10	13.474	11.726	8.188
	.07		20.805	19.58	15.474	13.326	9.188
	.08		23.605	22.06	15.474	14.926	10.188
	.09		26.405	24.54	19.474	16.526	11.188
	.10		29.205	27.02	21.474	18.126	12.188
	.11		32.005	29.50	23.474	19.726	13.188
	.12		34.805	31.98	25.474	21.326	14.188
	.13		37.605	34.46	27.474	22.926	15.188
	.14		40.405	36.94	29.474	24.526	16.188
	.15		43.205	39.42	31.474	26.126	17.188

The foregoing table shows comparative costs for obtaining 80 candle-power from Carbon, Gem, Tantalum and "Mazda" lamps. At the present electric power rates, a large saving is noted in favor of "Mazda" lamps. At a power rate of 10 cents per kilowatt hour the cost for 80 candle-power of "Mazda" light for 1000 hours' burning (which includes the lamp cost) is about three-sevenths of the actual cost of power alone for equal candle-power from carbon lamps of 3.5 w.p.c.

The chief objection to the tungsten filament lamp is its fragile nature, but this is certain to be overcome when we notice the same objections were made about the carbon lamp about fifteen years ago. I feel safe in predicting a much higher efficiency for the "Mazda" lamp and a perfecting of the strength of the filament to stand any reasonable amount of jar and rough usage.

DRY CELLS.

SAUL DUSHMAN, B.A.

At a meeting of the American Electro-Chemical Society in October, 1909, the following remarks were made by President Baekeland:

"The dry cell, modest as it may appear as a unit, forms the base of a more important industry than most of us imagine. The number of dry cells consumed yearly can be counted by millions, the industry is more important than many electro-chemical industries of which we hear much more in our meetings, and read more in print. It so happens, at the same time, that it is one of the subjects on which a very large amount of misinformation is abroad, misinformation sometimes sent out purposely, but most of the time involuntarily through pure ignorance. It is almost impossible to tell beforehand whether a dry cell is good or not, because here everything goes by experience. In fact, I would almost say that the test of a dry cell is to use it for a particular purpose. We are familiar with the phrase 'the test of the pudding is in the eating of it,' but the trouble is that after the eating of the pudding there is none left."

This statement, taken together with the fact that the annual production of dry cells in the United States alone is well over the 50,000,000 mark, shows the importance of a more perfect knowledge of the behavior of dry cells than is at present possessed by the great majority of users.

Construction of Dry Cells.

Probably 80 per cent. of the dry cells made and used are of the so-called No. 6 size, about 6 inches high and 2.5 inches in diameter. A cylindrical zinc container, insulated externally by cardboard, acts as negative pole, while the positive consists of a carbon rod or plate. The zinc can is lined with "an insulating absorbent layer, which serves as a reservoir for electrolyte and as a means of separating the two electrodes. This layer is usually made of a special grade of pulpboard manufactured for the purpose under conditions that prevent the presence of ingredients harmful to the cell." The filling material is packed in between the zinc container and the carbon electrode. A fairly typical filling-mixture has the following composition:

> 10 parts manganese dioxide (MnO_2),
> 10 parts carbon or graphite,
> 2 parts sal ammoniac (NH_4Cl),
> 1 part zinc chloride ($ZnCl_2$).

Water is added to make a heavy paste and to produce the requisite electrolyte.

Chemical Reactions.

When the cell is discharging the solution of $ZnCl_2$ and NH_4Cl is decomposed by the current; the chlorine liberated at the anode attacks the zinc and forms more $ZnCl_2$, while the hydrogen gas which would otherwise dissolve in the carbon and produce a back e.m.f. is "depolarized" by the MnO_2. The latter is reduced in this process to a manganous salt. The increase in the total amount of $ZnCl_2$ in the cell as well as the decrease in available MnO_2 lead to a gradual diminution in the electromotive force of the cell. The addition of the $ZnCl_2$ to the filling mixture might appear, therefore, to be a questionable procedure. Nevertheless, it has been stated that this addition is necessary.

As the rate of the chemical reactions at the electrodes is always proportional to the rate at which current is taken out of the cell (Faraday's Law) it is readily seen that fresh chemicals must continually replace those used up at the immediate surfaces of the electrodes.

If current is drawn out of the cell at a rate greater than that at which the fresh chemicals can diffuse in towards the electrodes, the electromotive force begins to drop very rapidly, and the cell is said to be "polarized." It follows that the larger the surface of the zinc and the more finely pulverized the manganese-dioxide carbon mixture, the greater the current that may be taken out of the cell without polarizing it. Furthermore, the more homogeneous the mixture, the more rapidly the cell will recover from the effects of heavy currents.

Voltage and Capacity.

The open-circuit voltage varies in different types from 1.5 to 1.6 volts, and is on the average 1.56 volts. The electrical energy output varies greatly with the manner in which the cell is used. When short-circuited an average cell will give from 18 to 25 amperes, falling at the end of one hour to 10 amps. at about 0.5 volt. It is, however, a serious mistake to consider this short-circuit current a measure of the value of a cell. As a matter of fact it often happens that cells which give a high short-circuit current fail badly when tested by the standard of service-capacity, that is, by the number of ampere-hours delivered when the cell is discharged through a resistance of 16 ohms. The following table, taken from Ordway's paper, illustrates this point:

TABLE I.
Short-circuit Current and Service Capacity of Dry Cells.

Brand of Cell.	Short-circuit Current.	Service Capacity.
A	33.0	24.0
B	29.5	46.0
D	24.5	33.5
G	22.5	40.0
J	20.9	18.2
L	20.2	36.0
Q	19.2	38.6
S	18.2	27.1
W	9.7	16.0

An important fact about dry cells is that the service-capacity, as well as the short-circuit current, drops even when the cell is standing on open circuit. This drop in the short-circuit current is known as the "shelf-life" in the trade, and although in itself the short-circuit current may be of no value as a measure of the efficiency of the cell, yet a considerable variation in this current given by an old cell and that given by a new cell is a sure indication of a serious fault in the mechanical construction of the cell.

The energy output of a dry cell depends upon the conditions under which the cell is discharged. The discharge may be either continuous or intermittent, and in the usual case it occurs through a constant resistance. The following table, also taken from Ordway's comprehensive paper, represents typical results obtained at the Research Laboratory of the National Carbon Co., Cleveland:

TABLE II.
Watt Hours from No. 6 Dry Cells Discharged Continuously.

End Point in Volts.	Resistance used in Ohms.						
	2	4	8	16	24	32	40
1.2	3.7	4.3	8.1	15.2	18.8	21.7	23.8
1.0	6.7	13.0	16.5	26.9	33.4	39.8	42.0
0.8	9.7	16.3	21.5	32.8	40.3	44.6	50.6
0.6	12.5	19.4	26.6	48.9	49.5	52.7	53.2
0.4	15.4	27.3	39.1	52.6	54.3	58.2	54.8
0.2	19.8	32.6	41.5	53.3	55.2	59.3	57.1

F. H. Loveridge gives the following values of the energy output of dry cells discharged to 0.5 volt:

With 1 ohm...................... 7 watt hours
With 2 ohms..................... 8 watt hours
With 4 ohms.....................14 watt hours
With 6 ohms.....................21 watt hours
With 10 ohms....................24 watt hours

When discharged intermittently through the same resistanecs, the cell gives a longer actual service as long as the voltage does not diminish to less than one-third its original value. After this stage is attained the cell can be operated more efficiently on a continuous discharge.

C. F. Burgess carried out a large number of tests with dry cells obtained from eleven different manufacturers. The number of watt hours delivered by the cells on intermittent service varied from 58.4 to 3.3, and the initial resistance from 0.68 to 0.17 ohm. Burgess states that a good dry cell, when used on intermittent service should deliver 2.27 watt hours per cubic inch contents, and should have a resistance not greater than 0.3 ohm.

The Effect of Temperature.

While the effect of increased temperature on the electromotive force of a dry cell is only slight, the effect on the shelf life is very deleterious. Cells which initially give a short-circuited current of 20 amperes will indicate a current of less than 1 ampere when stored for several months at a temperature of 50° C, while the same cells stored for an equal length of time at ordinary temperature will indicate 18 or 19 amperes. This accounts for the advice of manufacturers that dealers and users store their cells in a cool place.

"The service capacity of dry cells may be either increased or decreased by raising the temperature according to the conditions under which the cells are tested. In general, more service of a severe nature will be obtained if the cells be moderately warmed, while with very light service it is advantageous to keep the cells cool. It is dangerous, under any circumstances, to use dry cells at temperature much above 50°C."

Testing of Dry Cells.

The object of a test is to determine the value of any piece of apparatus or material for a certain specific purpose. Such a test may be either comparative or absolute. In the case of the dry cell the first test to be applied was the measurement of the short-circuit current. At the best this could only be a comparative test; but, as pointed out above, the conclusions to be drawn from it are useless as regards the service capacity. The continuous resistance test was therefore tried next. It was soon found, however, that here again the conclusions of such tests were not applicable to conditions in actual practice. While one cell may be superior to another when discharged continuously, yet the reverse may be the case when both cells are tried on the same ignition arrangement. Or two cells may show equal energy output on continuous discharge, but be unequally efficient when used on telephone service. Now the question which obviously interests the consumer is not the amount of energy which the cell is capable of delivering under certain standard conditions, but **the actual hours of service that he may expect to obtain when operating a definite piece of apparatus under definite conditions with the given cell.** The conclusion, therefore, at which both manufacturers and consumers have arrived is to devise tests that shall duplicate as nearly as possible the conditions under which the cells are to be used, thus making it possible to obtain data by which the cells can be rated for a given service.

As the two main uses of dry cells are for telephone and ignition service, we shall briefly review the nature of the tests which have been elaborated to determine the value of a dry cell for either purpose.

(1) Telephone Service.

" To represent as nearly as possible average telephone service, two or three dry cells should be connected in series through about 15 ohms for as many periods each day as there are calls on an average telephone, each period being the length of an average telephone conversation, and the cut-off point being taken as the lowest working voltage that will give satisfactory service."

A simple test may be carried out as follows: Three cells in series are discharged through 20 ohms resistance for two minutes per hour, 24 hours per day, until the working voltage (taken just before the end of a contact) decreases to 2.8 volts. The test may be varied by discharging three cells in series through 15 or 20 ohms for four minutes per hour, 10 hours per day, until the same voltage is attained.

(2) Ignition Service.

" In the usual type of ignition apparatus the battery current passes through the primary of an induction coil at the right instant in the engine's cycle, and by means of a vibrator is made to rise and fall very rapidly, thus producing pulsations of very high voltage in the secondary, and a series of sparks at the spark plug which ignites the explosive mixture in the engine cylinder. The primary of the average induction coil has a resistance of about 0.5 ohm, and if the circuit through the coil were to be closed indefinitely the current would rise to from 12 to 16 amperes, depending on the voltage of the battery used. The action of the vibrator is so rapid, however, that the current does not have time to build up to its maximum value, and the actual initial impulse rises to only three or four amperes." Tests in which actual induction coils were used have been found to be useless owing to the fact that it is almost impossible to keep the coil in permanent adjustment. The following is given by Ordway as a standard test for ignition-cells: "Six cells in series are discharged through 16 ohms resistance for two hours per day, one hour in the morning and one hour in the afternoon. About twice a week the cells are short circuited through a half-ohm coil for a few seconds, and the current flowing is noted. When this current flowing, or impulse, reaches four amperes, the battery is removed from the test, and the life of the battery is reported as the hours of service given under the above conditions to four amperes impulse through a half-ohm coil. A modern dry cell of good quality will give 30 hours of service on this test at ordinary room temperature of, say, 22 to 24° Centigrade."

Conclusion.

It is evident from the above remarks that there is no single test which the manufacturer or consumer can apply to a dry cell to indicate quickly and simply its value. Each application requires special standards of its own, and it is not at all reasonable to expect that the same type of cell will fulfil the requirements of different users.

List of References.

J. W. Brown. The Use of Dry Cells on Ignition Service. Trans. Am. Electrochem. Soc., 13, 172-185.

C. F. Burgess and C. Hambuechen. Certain Characteristics of Dry Cells. Trans. Am. Electrochem. Soc., 16, 97-108.

F. H. Loveridge. Dry Cell Tests. Trans. Am. Electrochem. Soc., 16, 109-124.

D. L. Ordway. Some Characteristics of the Modern Dry Cell. Trans. Am. Electrochem. Soc., 17, 341-366. All the passages quoted above are taken from this paper.

C. F. Burgess. Tests of Dry Cells. Electr. World, 39, 156.

ARCHITECTURAL DESIGN.

The illustrations used in this article are intended to represent the work done by the different classes in the Department of Architecture of the Faculty of Applied Science. Some problems are given to illustrate the nature of the requirements, and the wide scope for the students' original ideas.

The department is in the charge of Mr. A. W. McConnell, who is a graduate of this department of the University, being a member of the class of '06. Mr. McConnell has further spent several years studying design in Paris and has brought back with him a thorough knowledge of the methods of teaching design as practised in the Atelier de M. Gromort.

A large number of the leading architects of this and other cities have visited the Department of Architecture during the present session and have expressed unanimously their approval of the work done by the students in this department. Those who are in the position to know, compare favorably the work in Toronto University with that of similar departments of American universities.

Applied Science understands the encouragement that such appreciation has for the students in architecture, and hopes other members of the profession will pay visits of inspection to their studios in the Engineering Building.

Following is a list of problems that are typical of those comprising the course:—

First Year Problems.

I. Survey of Building.

The student is asked to make measurements of a building or part of a building in his regular course of surveying and represent the same in plan, section and elevation in his studio. (See Plates II. and III.)

II. Orders of Architecture.

Arrangement of plates in it, using the various orders of architecture. (See Plates IV. V. and VI.)

Second Year Problems.

I. Porch for City Residence.
(See Plate VII.)

II. A Small Public Library.

The student is asked to make sketch plans in the class-room, and to work out the problem in plan, section and elevation in the studio, developing his original idea. (See Plate VIII.)

Third Year Problems.

I. A Dispensary for Children.

(See Plate IX.)

II. A Railway Station for Suburban Line.

This building is to be erected in a suburb of a large city facing a bridge over a viaduct on a busy thoroughfare. The line passes 25 feet below the street level. The building will not exceed 80 feet frontage.

The following accommodation is required:—

    (a) Large hall.
    (b) Two ticket offices.
    (c) Station master.
    (d) Baggage rooms with elevators for both lines.
    (e) Newspapers, etc.

(See Plate X.)

III. A Small City Hall.

(See Plate XI.)

IV. Music Hall.

This building is to be erected in a public park in a city of 40,000. The front elevation is not to exceed 80 feet in length. Concerts will be given day and night.

Required—

    (a) Vestibule with ticket offices.
    (b) Porticos, also to be used as a shelter for the people in the park.
    (c) Auditorium with seating space for 400.
    (d) Stage for 50 musicians.
    (e) Library for music.
    (f) Lobby for artists.
    (g) W. C., etc.

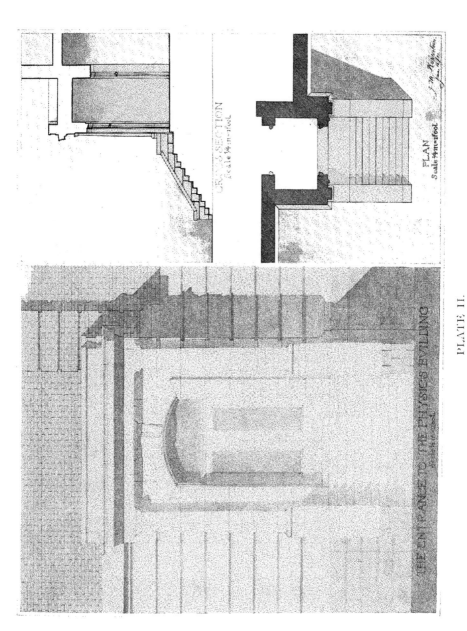

CROSS SECTION
Scale ¼in=1foot.

PLAN
Scale ¼in=1foot.

THE ENTRANCE TO THE PHYSICS BUILDING

PLATE II.

Measured Drawing by J. M. Robertson.

(See First Year Problem I.)

· SVRVEY · OF · BVILDING ·

· OBSERVATORY ·
· UNIVERSITY · OF · TORONTO ·

PLATE III.
Measured Drawing by R. S. McConnell, '13·
(See First Year Problem I.)

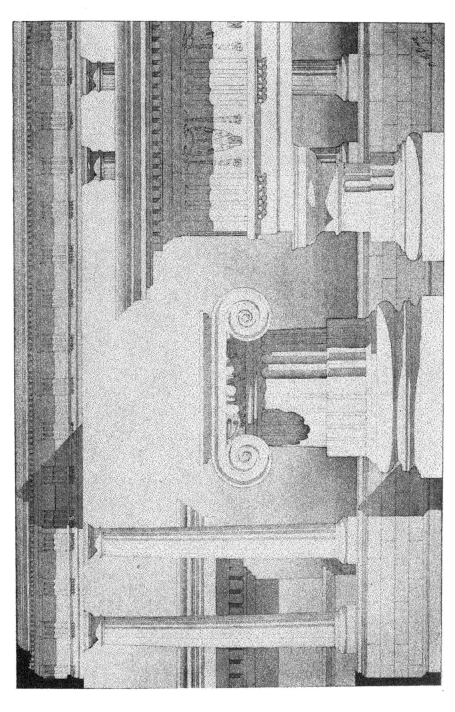

PLATE IV.

PLATE V.
Designed by R S. McConnell, '13.
(See First Year Problem II.)

ROMAN DORIC ORDER   VIGNOLA

PLATE VI.
Designed by R. S. McConnell, '13.
(See First Year Problem II.)

Porch for City Residence.

PLATE VII.
Designed by H. H. Madill, '11·

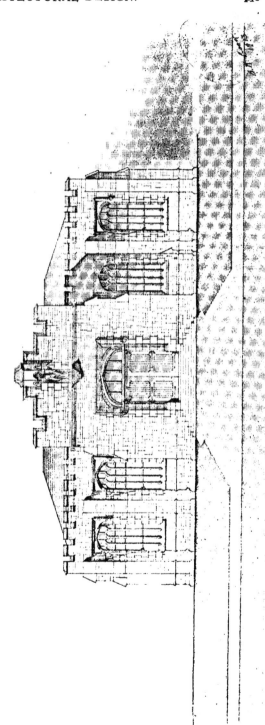

A Small Public Library.

PLATE VIII. Designed by B. R. Coon.
(See Second Year Problem II.)

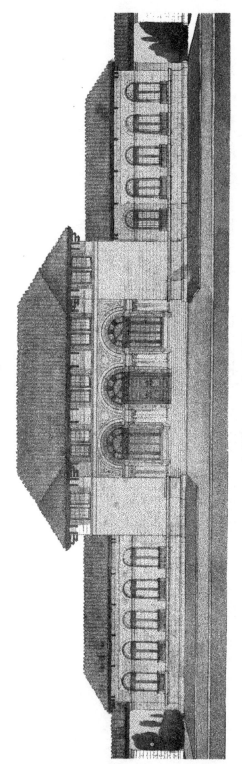

A Dispensary for Children.
PLATE IX.
Designed by J. B. K. Fisken.

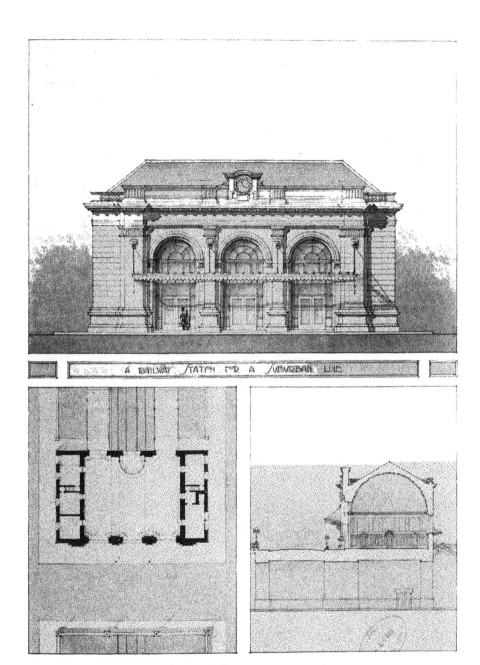

A Railway Station for Suburban Line:
PLATE X.
Designed by H. H. Madill, '11.
(See Third Year Problem II.

A Small City Hall.

PLATE XI.

Designed by T. L. Rowe.

PLATE XII. Music Hall.
Designed by J. B. K. Fisken.
(See Third Year Problem IV.)

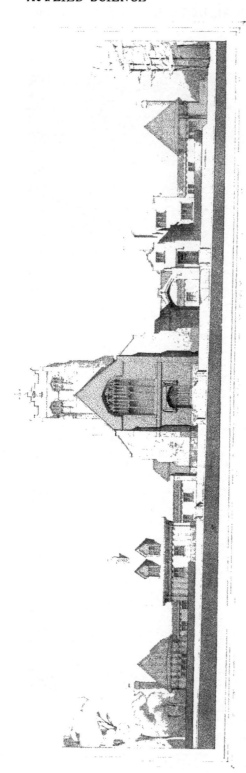

A Home of Rest for the Aged Poor, with detached Chapel and Manse.

PLATE XIII.

Designed by H. H. Madill, '11.

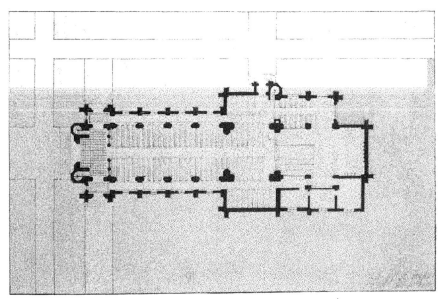

PLATE XIV. · A Gothic Church.
Designed by J. B. K. Fisken.

PLATE XV.　　A Gothic Church.
Designed by J. B. K. Fisken.

THE ENGINEERING SOCIETY DINNER

The twenty-second annual banquet of the Engineering Society has been added to its list of predecessors as a success from every point of view, and the executive upon whose shoulders the responsibility was placed are reaping a reward of praise for their untiring efforts.

To undergraduates the dinner has been a yearly opportunity to assemble as one of the strongest organizations of its kind, to meet the graduates of previous years, and to listen to men who, as statesmen, are at home when in the midst of plans and conceptions for the material betterment of Canada; and who endeavor to impress upon the undergraduates that it is they, as engineers, in whose hands the carrying out of these plans will be placed.

To graduates, the banquet is a reunion, a rendezvous of old classmates and instructors, and the founding of an acquaintance with those who are preparing to assist them in the building of productive transportation and commercial economy, and in the managing of material affairs in general.

To those who come as the guests for the evening, the dinner gives an idea of the magnitude of this Faculty of the University, of the student " esprit de corps " that is so necessary to its advancement, and of the forces behind it all, the forces to which many of the most successful engineers in America claim they owe a great deal and feel proud of, and on which the student to-day relies so much.

What is being accomplished in the Faculty of Applied Science and Engineering has been regrettably unknown to the majority of statesmen, unless they happen to be engineers, or have been in contact with the graduates. But the methods employed, the equipment used, and the needs of both these factors, to operate on the ever-increasing scale cannot be brought to their attention unless they become interested in the profession of engineering in general. The annual banquet of the Engineering Society accomplishes a very great deal toward this end.

Upwards of six hundred were present on the evening of January 19th in Convocation Hall. Mr. A. D. Campbell, the President of the Society, officiated as toastmaster. In welcoming the members of the Toronto Board of Trade as guests of the evening he dwelt briefly upon the fact that some of their interests ran along the same or parallel lines with many of our own future interests. He extended to all, on behalf of the Faculty and its students, an invitation to visit our buildings and equipment, and to study what is here being accomplished for and by young men.

Mr. L. E. Jones, '11, was called upon to propose the toast to " Canada," and coupled with it the names of Dr. Robertson,

The Twenty-second Annual Dinner.

Chairman of the Royal Commission on Technical Education, and Major Leonard, of the Board of Governors. In reply Dr. Robertson spoke as follows:

Dr. Robertson.

Mr. President, Dr. Falconer and Gentlemen,—I thank you, on behalf of the Royal Commission on Industrial Training and Technical Education, for the honor and pleasure of being one of your guests to-night and, for myself, for the honor and privilege of speaking briefly to the students in the Faculty of Applied Science in such a great institution as the University of Toronto.

Canada is becoming a great country, while still a comparatively young country. We older men like to be associated with youth. That helps to keep us young in outlook and in faith. The young in heart, the clear in mind with trained abilities and spirits animated by goodwill, are needed for the upbuilding of Canada. Here you have a good place for training. Teaching is not the whole of training, and telling is only a small part of teaching. The other day in New York I had a talk with two men eminent for their labors and gifts towards the improvement of technical and professional education. Referring to the recent bulletin issued by the Carnegie foundation it was confirmed that the University of Toronto stands in the very front in its physics and engineering departments, and that its splendid equipment was cared for and used with an efficiency not surpassed by any other institution which prepares men for the profession of engineers. Canada is a big place with great possibilities. It needs big men trained to cope with them and make the most of them for the benefit of all the people. The big tasks cannot be undertaken with hope of good results unless men and women of native talent are trained for the work."

In referring to the enormous extent of Canada and its vast, varied and valuable natural resources, Dr. Robertson said that conservation of resources does not mean keeping them out of use. It calls for their development by the application of scientific methods directed by capable men. That makes it possible not merely to pass on to our successors the natural resources unimpaired and undiminished by waste, but actually improved and extended by intelligent management. Competent engineers are needed more and more for all that.

Dr. Robertson gave an outline of the work of the Commission on Industrial Education. Its mission was to enquire, to gather information, and not to propose policies or advocate systems. In its survey of Canada it had already heard the testimony of some thirteen hundred men and women; it had among its witnesses such men as Dr. Falconer, the honored and beloved President of the University, and also men and women who worked at benches and on machines for daily wages. There was agreement in the desire expressed from all sides that Canadians should be industrially efficient; and efficient for all the duties which came to them as workers, as citizens, and as men and women. Trained leaders are no less required than skilful workers. Intelligence, practical ability and goodwill, are needed all through the population, in the ranks with the staff and most of all at "headquarters."

After referring to the production of wealth from natural resources by farming, mining, fishing, lumbering and manufacturing. Dr. Robertson spoke of the essential partnerships which exist between the dieffrent localities and provinces of the Dominion. "We have seen nothing in our survey of this country which indicated that the progress and prosperity of auy locality

Dr. J. W. Robertson.

was made to the injury of the others. We are partners as provinces and partners in serving Canada in the various occupations which we follow. Men on the football field strive for the supremacy, for winning in the game. But members of competing teams are not hostile to each other. They are friends and supporters of the game. Team play is the best play for the individual as well as for the team. In the national game of playing for the greatness and goodness of Canada among the nations the same holds good. We must inspire the newcomers, the stream of foreign blood pouring into our citizenship, with the same spirit. They are not only among us, they are of us. Our safety and their welfare are not to be ensured by aloofness or separations or disparagements, but by helping them up to our standards of fair play, by playing the game that way ourselves.

Particularly by means of education we must bring them up to our, plane and aim higher ourselves. We must teach them to love and serve Canada, and example is here stronger than precept. You young men particularly, who are among the elect in opportunity, as shown by your attendance at this University, must set an example of clean living and clear thinking, and hard work with good will. Thus, you can extend the blessings of your heritage of blood and opportunity for the glory of Canada, and the good of your fellow-citizens throughout the whole land."

Major Leonard spoke briefly, basing his remarks upon the amazing advance Canada has made during the past decade. He emphasized Dr. Robertson's advice to undergraduates to endeavor to realize the responsibility that rested with them.

"Our engineers must be trained for the work they must do. They must be trained to meet the problems of transportation, for instance. We already have great railroads and canals. We are building greater railroads and greater canals.

" The passing generations have called to our country many aliens who are now living here with us. The coming generation of engineers will employ many of these aliens, and you men must bring their ideals up to the high plane of your own. Your duties are greater than the duties of those who have built the great railroads and canals, and brought these aliens into our country. Many of these men have come from parts of Europe where they know nothing of responsibility to government such as we know. It is for you to uplift their ideals.

" I was very glad to hear Mr. Jones refer to military training, and of the manner in which the sentiment was received. There is no place where one can acquire executive ability better than in military training, the training whereby a handful of men, such as our Mounted Police, can maintain law and order in such a vast territory as our Northwest, or the preserving of order in India, where a few British troops are surrounded by hordes of semi-hostile people. Such things are possible only by discipline, the discipline which is maintained throughout the entire British army and navy. This is the best school to teach it, and it will be of service to you throughout your whole life."

Mr. R. A. Sara, '09, then proposed the toast, "Canadian Industries," and Mr. R. S. Gourlay, the President of the Toronto Board of Trade, was called upon to reply to this toast.

Mr. R. S. Gourlay.

"You have asked me to respond to the toast of 'Canadian Industries,' and at the outset I would compliment you on the term used. It is not as we hear it so frequently in tariff discussions, 'a few manufacturers,' but Canadian Industries, which cover an area from the Atlantic to the Pacific, which I cannot illustrate better than to give you a picture of my experiences last summer in Kelowna, B.C., where the first manufacturing plant was started last year by a man who had hitherto been simply a grower, and also a little village of 125 inhabitants on the Digby Neck, N.S., which is typical of the whole of Canada. I found there a contented happy settlement, its people engaged in agriculture and fisheries, and the product of the fishery being cured, smoked and packed for the market, even to the making of the cans, whilst, in addition, a buyer purchased another part

Mr. R. S. Gourlay.

of the fishing by-product for the export to the United States for manufacture into isinglass, because we have, as yet, no manufacture of this commodity in Canada.

"This, I say, is a picture of the whole of Canada. During the past few years industries have and are being established everywhere so that we now employ in Canada some 445,000 people in these industries, and if we allow a modest estimate for those dependent on them we have easily a million people who are directly supported from these industries, at least one-eighth of our population, without numbering at all the professional, commercial and transportation classes, who also derive the larger part of their incomes from these industries, indirectly it is true, but none the less surely.

"Another thought I would present is that though our industries are making progress, the output last year being estimated at a billion dollars, yet we are as yet only on the highway to being an industrial nation, as our customs imports show us that we imported some $300,000,000 worth of manufactured articles last year, much of which is sold in competition with the same class of goods made in Canada, and the balance because, as yet, we have not entered upon their manufacture.

" For instance, Canada has deposits of silica, yet we make no plate-glass, no window glass and no glass blocks for cutting. Our glass manufactures cover only a fraction of our importation and are confined to bottles, fruit jars, chimneys, globes, and tumblers, with practically none of the finer glass products.

" Again, consider our immense deposits of clay everywhere, and yet beyond building material we only make brick and the common classes of tiles, flower pots, and earthenware, nothing of such table and artistic ware as we use every day, and without which our homes would look poverty stricken if they had to use only made-in-Canada clay products.

"Truly in these classes of industries many lines have not yet reached ' the infant stage,' and so is it in other directions, and there is, therefore, much room for you, young men, who are to go out as leaders and experts in industrial life, to apply your knowledge and see that we make even more rapid progress in the way to becoming a fully developed industrial nation.

" Still another thought—the Government statistics indicate that the relation in Canada between workmen and master is more ideal than elsewhere, in the matter of wages, for the Government statistics show that in Canada much higher wages are paid than in Great Britain and Germany, and wages fully as high are paid in Canada as in the United States, where by the use of larger plants and greater development in specialization the output for the same wage is from 20 to 25 per cent. greater per workman. Whilst a still further indication of this happy, ideal condition between the master and men is revealed in the Government labor reports as to strike conditions. Last year with 445,000 workers there were, small and great, only 68 labor disturbances, affecting 17,000 employees, and that from these if you count out 12,500 employed in mining, building trades and transportation companies, there were but 4,500 people interested in all industries from fishing to street laborers, who, even for a period of a day or two were in any labor trouble, just 1 per cent. of the working force, a record such as cannot be found in any other country under the sun.

" In a word, you are entering on your life labor at a time when Canada is fairly on the highway to being, if not checked, a fully developed industrial nation, with conditions that are at present more ideal than elsewhere, when between master and man there is that spirit of co-partnership that more than aught else will make this country a great industrial nation.

" It is for you, young men, to rightly apprehend and cultivate this spirit, that combined with your skill in specialization and expert trained guidance will so increase the output, augment the workman's wage and the employer's profit as to develop still further this ideal condition.

"I have but one more thought. I am old enough to have begun business at the time when I had to choose between remaining in Canada with limited prospects or to leave for a field of larger possibilities, and I stayed, when the large number of my boyhood friends were finding work to the south of us. The emigration from Canada was enormous, young men of to-day cannot comprehend how enormous, and I also have lived through a period when Canadian industries were so few and far between that Canadian products in my line had to be sold with a guarantee that if they did not please on trial, they would be replaced for the purchaser with American products. But we are past that point, and now the Canadian product is recognized in all parts of the world, as of a class that in point of merit averages a standard that is not excelled in any other country.

" Young men, this fruitage of past years is your heritage, prize it highly, don't part with it, or do aught to lose it. It is your birthright, don't for any temporary advantage become an Esau and sell it for 'a mess of pottage.' "

President Campbell then called upon Mr. L. R. Wilson, '09, to propose the toast " The Legislature." In doing so Mr. Wilson reviewed the interest the Legislative Assembly had taken in the engineering progress of the country, and in endeavoring to further technical education. He made mention of the splendid work the different departments of the Legislature, especially that of Public Works and in the Department of Agriculture.

The Hon. J. S. Duff, in response, expressed his appreciation on the subject of military training for the Canadian boy, and also voiced his intention of watching with interest the great progress of the University, especially of the Faculty of Applied Science.

" The University " was proposed by J. Galbraith, Jr., who likened its success to that of its football teams, and pronounced " team work " as being the mainstay of success in both. He voiced the appreciation of the student body with regard to the magnificent gift of buildings lately presented by the Massey estate.

President Falconer, upon rising to respond, was received with the usual enthusiasm, which bespeaks the place he holds in the hearts of the undergraduates of the Faculty of Applied Science.

President Falconer.

"It was an excellent idea of the Engineering Society to invite here to-night Dr. Robertson, also the members of the Toronto Board of Trade, because many of your will go out and engage in the industries represented by these gentlemen. You engineers must more and more tend to bridge the gap existing between educated men and the class that have often thought themselves to be the unprivileged class, because the privilege of education has been denied to them. You men will come in contact with workmen, and it is through you that some of the benefits of education will reach them. These benefits will enter into the finished product, and also into the life of the man while the product is being brought to its completion.

" Does the engineer work only for the completion of any building or any great work? Ask your graduates in New York whether they sit still and contemplate the building when it is finished. Is that their attitude? By no means. As soon as one piece of work is done they look for another. Enjoyment comes not mainly in the completion but in the process. Day by day you will be putting into the workman the desire to do his work with all his intelligence, and you will be adding to the contentment of his life and to your own happiness. This is the secret of industrial efficiency of the highest sort. To set a high ideal and to realize the ideal is what we aim at in the University and in the Faculty of Applied Science. You are trained to become engineers, but beyond this you must put into the engineering work the purpose of your life, and this cannot be fully reached apart from the development of those who with you carry your engineering designs into execution."

In replying to the toast " The Faculty," proposed by Mr. Chas. Webster, Dean Galbraith commanded the rapt attention of every hearer as he reminiscently reviewed the progress of the " School " since the days, thirty years ago, when Dr. Ellis and himself constituted the teaching staff, and the total enrolment consisted of seven students.

" It would delight the hearts of the founders of this institution and of those who taught here with me first, to see the class which numbered seven when we commenced, grown to nearly eight hundred.

" In the olden days the teaching was done by men from the Colleges. They were well up in their subjects but they had not been out in the world and lacked the contact with actual conditions. One of the great changes has been in the type of men on the faculty. We are cut off from the outside here, and one of the strongest tendencies as a university becomes more immense is to sever the bonds to a greater extent still. It is the duty of our graduates to bring back to us the fruits of their experience in the world with the learning gained here, and to keep us in touch with the outside."

The Dean referred briefly to his recent visit to Cobalt as a guest at the S. P. S. Temiskaming dinner, and praised the work done there by students.

"We are not afraid of lack of success for our graduates, and we are sure of their ability to carry on the work for which they were trained."

Before rising Dean Galbraith was greeted with a chorus from the undergraduates as follows:

"Long live our Dean!
Don't you hear them cheering?
Don't you hear them shouting as the Dean goes by?
Long live the Dean! That's the song we sing,
God bless our Dean! is the students' loving cry."

Mr. W. A. O'Flynn proposed "The Engineering Profession," and coupled with it the name of an old and most successful graduate, W. J. Francis, C.E., of Montreal. The address of Mr. Francis, as that of one student to others, appealed strongly to every student and young graduate present, and could not have been replaced by a more fitting or more appreciated response.

After some humorisms, referring more particularly to school affairs, he quoted Tredgold's historic definition of the Civil Engineer as one who directs the great sources of power in nature to the use and convenience of man. Measured by that standard, the man who employs the electric current to move the wheels of commerce is a civil engineer in the field of electricity; the man who sinks a shaft or drives a tunnel is a civil engineer in the art of mining. The name and the work it implies are most comprehensive and inclusive in their significance. Expressing his great satisfaction that the civil engineer is no longer considered a sort of glorified plumber by the public and that the profession is rapidly taking its place beside those other professions of law, medicine and theology, the speaker made the statement that the reason this condition was so much delayed lay largely with the engineers themselves.

He deplored the practice of many engineers in lobbying for work, and in countenancing the action of individuals and corporations in need of the services of an engineer, who are seeking the lowest bidder regardless of qualifications or reputation. As a parallel between engineers and their clients, he pictured a number of doctors bidding for a surgical operation and the patient in the throes of appendicitis waiting the results of the tenders in order to determine who would do the carving for the least number of dollars.

Proceeding, he spoke of the Canadian Society of Civil Engineers as probably the "greatest professional organization in Canada to-day," and as one which is doing excellent work in elevating the status of the engineering profession, and in en-

Mr. W. J. Francis, C.E.

deavoring to create of standard of engineering ethics in Canada. That society, he said, had signally honored the University of Toronto. Dean Galbraith is a past president, and Professor Haultain and many of the other graduates are councillors. A good number of the graduates are members, and the others should all be connected with the Society. Mr. Francis urged the students to affiliate themselves with the Canadian Society of Civil Engineers and so assist in the establishment of a high standard of professional conduct.

He referred reminiscently to his student days at Toronto twenty odd years ago. when the staff consisted of Dean Galbraith, Dr. Coleman, and Dr. Ellis of the present staff, and a few others who have since crossed the Great Divide. "We didn't call him the Dean then, we called him Professor—when he was there; you know what we called him when he wasn't there." Those were the days of "the little red schoolhouse" for the engineering student. In the years that have passed since then, great expansion, at that time not even dreamed of, has taken place, but "Five o'clock, gentlemen!" still resounds through the old building. In his closing remarks, he urged the students to master the fundamentals, and to be thorough in their work. He told them they had chosen for their life work that profession which had more to do with the world's wonderful advancement than any other, and advised them to adhere to that straight and undeviating line of conduct which should at all times characterize an engineer. By doing these things he felt that in future years they would reflect credit on their Alma Mater, which he regarded as one of the "greatest institutions in the Dominion."

The lengthy programme of speeches was relieved by musical selections, rendered in a most pleasing style by the Science Octette, the members of which this year are, Messrs. W. A. O'Flynn, W. A. Costain, P. S. McLean, G. B. Macaulay, J. H. Craig, W. C. Blackwood, R. B. Chandler, G. J. Mickler. The Octette is again under the leadership of Clayton E. Bush.

The success of the twenty-second annual dinner is due to the united efforts of the entire executive, and to the able assistance of Prof. Wright and others of the staff.

———

W. T. Main, '93, is division engineer, on the Wisconsin division of the C. & N. W. Ry.

W. G. McGeorge, '08, is conducting a general engineering practice in Chatham, Ont.

A. T. C. McMaster, '01, is engineer on design and construction of basic converter and reveberatory plants, The Canadian Copper Co., Copper Cliff, Ontario.

J. L. Morris, '81, and W. J. Moore, '06, are in partnership at Pembroke, Ont, under the name of Morris and Moore, surveyors and general engineers.

A. E. Pickering, '04, is with the Lake Superior Power Co., as superintendent.

R. S. Smart, '04, is in Ottawa, as manager Fetherstonhaugh & Co., Patent Solicitors.

P. H. Stock, '09, is resident engineer, for the Niagara, St. Catharines and Toronto Ry.

W. V. Taylor, '93, is Harbour Commissioner for the city of Quebec.

A. F. Wells, '04, is president of the firm of Wells and Gray, Ltd., engineers and contractors, Toronto.

APPLIED SCIENCE

INCORPORATED WITH

Transactions of the University of Toronto Engineering Society

DEVOTED TO THE INTERESTS OF ENGINEERING,. ARCHITECTURE
AND APPLIED CHEMISTRY AT THE UNIVERSITY OF TORONTO.

Published monthly during the College year by the University of Toronto Engineering Society

EDITORIAL

LIST OF GRADUATES WHOSE ADDRESSES ARE UNKNOWN.

The endeavor to have as complete a list as possible of the present addresses of graduates will be greatly aided by any information that may come from readers of Applied Science who are acquainted with the whereabouts of any of these men.

1882—D. Jeffrey, J. H. Kennedy.
1885—E. E. Henderson.
1887—A. E. Lott, F. Martin.
1888—D. B. Brown, K. Rose.
1892—J. R. Allan.
1893—W. Mines.
1894—H. F. Barker, A. L. McTaggart.
1895—W. M. Brodie, H. S. Hull.
1897—W. R. Smiley, E. A. Weldon.

1898—J. E. Lavrock, F. W. McNaughton, A. E. Shipley.

1899—G. A. Clothier, C. Cooper, J. C. Elliott, W. E. Foreman, E. Guy.

1900—J. Clark, J. E. Davison, H. A. Dixon, H. S. Holcroft, J. C. Johnston, R. E. McArthur.

1901—J. L. R. Parsons.

1902—J. M. Brown, A. R. Campbell, F. T. Conlon, H. V. Connor, A. C. Goodwin, J. T. Mackay, R. S. Mennie.

1903—H. G. Acres, R. E. George, W. A. Gourlay, J. A. Horton, A. L. McNaughton, C. A. Maus, M. L. Miller.

1904—P. C. Coates, O. B. McCuaig, G. G. McEwen, E. E. Moore, H. M. Weir.

1905—D. W. McKenzie, E. D. O'Brien, E. P. A. Phillips, C. H. Shirriff, W. F. Stubbs.

1906—C. Johnston, N. R. Robertson.

1907—A. P. Augustine, G. H. Broughton, R. A. Hare, E. W. Kay, D. J. McGugan, F. W. McNeil, J. D. Murray, A. F. Wilson.

1908—R. H. Douglas, C. Flint, W. A. Robinson, J. J. Stock.

BOOK REVIEWS.

Principles of Metallurgy.—Fulton.

(McGraw-Hall Book Company, New York—528 pages 6 ins x 9 ins. Illustrated.)

In this age of advancement, it is particularly noticeable that the conditions which have brought about the present tendency toward specialization have, of necessity, forced a similar state of affairs on the scientific literature of the day. It is only at long intervals that some author has the clear-sighted perception which enables him to compile a work in such a manner as to be extremely valuable to the practical reader who wishes far-reaching fundamental principles rather than minute descriptions of some one department.

That Professor Fulton has foreseen a long-felt want is quite evident when the scope of the above book is examined.

The writer has paid a great deal of attention to the very important subject of metallic alloys. In the chapter on "Physical Mixtures and Thermal Analysis," cooling curves and freezing point curves are very clearly discussed. From this, the author goes on to point out the various types of Binary freezing point curves. The pages devoted to Ternary systems might well have been added to. This part of the subject of metallic alloys is one which, above all, causes the student the most trouble: and it is little to be wondered at when one takes into consideration the manner in which information on the subject is scattered throughout the scientific journals.

Further on, various alloys are discussed, among which are the copper-zinc, copper-aluminum, copper-tin series. The diagrams for these alloys are plainly shown in full-page size. These diagrams, when taken with the discussion of the commercially most important mixtures, furnish a large amount of valuable information. The iron-carbon series has been condensed in a manner slightly out of keeping with its importance, although the writer seems to have recognized this by referring the reader to "works on metallography of iron and steel." A few of these references might well have been quoted, for although some references are given, yet they can hardly be said to be suitable for the average reader.

In chapter VI., the measurement of high temperature is taken up in an interesting manner, which, to say the least, robs a large subject of the usual mass of detail both wearisome and so often needless.

The chapter on "Slag," as pointed out in the preface, is based largely on the work of J. H. L. Vogt. The amount of condensed material in this section of the book is enormous. The mineralogical nature of slag, cooling curves, formation temperatures, specific heat and latent heat of fusion of slag are subjects which are briefly but ably set forth. The tables given in connection with these headings are comprehensive and clear.

Passing on to the next chapter, there are thirty-six pages devoted to mattes, bullion and splise. The various sulphide systems are illustrated by means of freezing point curves. In pointing out the characteristics of these systems, a large number of practical details are given which add to the value of the discussion.

From chapter XIII. one gets a very good idea of the various types of furnaces used in metallurgical work. The writer touches on the electric furnace. It might have been advisable to give more examples of furnaces used in this extremely important and growing branch of the industry.

. Besides the chapters specially mentioned are those on Physical Properties of Metals, Typical Metallurgical Operations, Refractory Materials for Furnaces, Fuels, Cumbustion, and a final section devoted to illustrating the physics and chemistry of a smelting operation.

One very good feature of the book is the large number of references given, some suitable for the average, and others for the advanced reader. The illustrations throughout are very clear. Taken as a whole, the book is a valuable addition to metallurgical literature.

(Reviewed by T. R. Loudon, B.A.,Sc., Dept. of Metallurgy.)

Applied Science

INCORPORATED WITH

TRANSACTIONS OF THE UNIVERSITY OF TORONTO ENGINEERING SOCIETY

Old Series Vol. 23　　　　MARCH, 1911　　　New Series Vol. IV. No. 5

A FEW POINTS ON REINFORCED CONCRETE DESIGN.

C. S. L. HERTZBERG, '05

In designing reinforced concrete structures, one is continually meeting minor problems upon which very little satisfactory information can be obtained from the numerous treatises on the subject. In the following paper I shall endeavor to enumerate a few points in design which are sometimes apparently not given the attention they deserve.

Footings have probably given more trouble to the designer, the erecting contractor and the owner than anything else in connection with reinforced concrete. Unequal settlement in footings is responsible for numerous unsightly deformities and cracks and some collapses.

The common type of reinforced concrete column footings is, of course, easily dealt with, and differs from a plain concrete footing only in its being designated as a flat slab to resist bending, instead of being sloped off as a pedestal. In this type of footing the centre of pressure coincides with the centre of gravity of the footing area, and the required size is formed directly from the load to be carried and the resisting power per square foot of the soil. Trouble is sometimes caused by having a footing too large in comparison with the size of the other footings in the same building. This is particularly liable to happen in the design of wall column footings in the following manner:

If the footings are designed to carry the total dead and live load, figuring each flooring of the building fully loaded, then the interior footings will, under probably loading, not stress the soil as highly as will the wall column footings. The reason for this is, of course, that the load figured to come on the wall column footings is usually about 70 per cent. dead load (which is present under all conditions), and 30 per cent. live load (which is never all there), while that figured on the interior column footings is generally about 40 per cent. dead and 60 per cent. live. As the live load on the footings of a building of five storeys or more is never more than 50 per cent. of the total live load, it will readily be seen that the pressure per square foot is less on interior footings than on exterior ones.

As all soil is compressed under any loading, the interior footings will not settle as much as the exterior ones, and the result is sometimes the cracking of floor beams and slabs.

The difficulty is overcome, to a certain extent, by the cus-

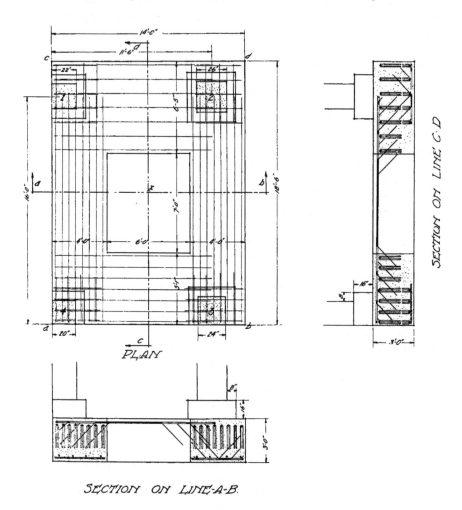

PLAN

SECTION ON LINE-A-B.

FIG. 1

tem of reducing the live load by about 50 per cent. in buildings of over a certain number of storeys. This, however, would appear to be insufficient, and it would seem that either a greater reduction should be figured in designing interior column footings, or else no reduction at all should be allowed in figuring exterior footings. It would also appear to be wise to even add a small percentage on corner column footings, as a much larger portion of the wall coming on these is dead weight.

It very often occurs that the footings under wall columns cannot be built to extend beyond the outside line of the column. In cases of this kind some sort of combination footing should be used. This is sometimes done by carrying the column in question on a cantilever beam, pinned down at the other end by one of the other columns. Care must be taken in this type to reduce the footing under the second column in proportion to the upward thrust from the end of cantilever beam.

A simpler method of treating the above is as follows:

Consider the wall column in question and the nearest interior column as acting together on a combined footing. Figure the loads coming on both columns and find the position of their resultant load. Add the two loads and divide by the soil value per square foot. This will give the required footing area. Design a footing of this area and varying in width from one end to the other in such a way that the centre of gravity of the area will coincide with the point of application of the resultant from the two column loads. The thickness of the footing and the reinforcing material must now be designed, treating the footing as an inverted beam, supported at the two columns and resisting the upward pressure of the soil, which will be of an intensity per square foot equal to the soil value minus the weight per square foot of the concrete in the footing.

The above method can be used for designing combination footings for any number of columns.

Figure I. shows a footing of this type designed to carry the four columns indicated, whose loads were: 1: 267,000 pounds; 2: 347,000 pounds; 3: 284,000 pounds, and 4: 197,000 pounds. The soil value assumed was 5,000 pounds per square foot. Column 1 was a corner column, 2 and 3 were wall columns, and 4 was an interior column. The footings could not extend beyond the lines A. B. and A. C. The footing was designed as follows:

Sum of column loads = 109,500 pounds.

Sum of moments about side a.b. = 10,272,166 foot pounds.

Therefore centre of pressure is $\dfrac{10272166}{109500} = 9'-4\frac{1}{2}''$ from a.b.

Taking moments about line a. c. we find centre of pressure is 7 ft. 0 in. a.c.

This locates the point x, the centre of pressure.

Area of footing required $= \dfrac{109500}{5000} = 219$ sq. ft.

The lengths, 18 ft. 6 in. and 10 ft. 0 in. of the sides a.c. and a .b. are now arbitrarily assumed.

Area of rectangle a. b. c. d. = 259 sq. ft.

Area of footing required = 219 sq. ft.

Area to be deducted = 40 sq. ft.

Deduct area, E.f.g.h. 7 x 6 = 42 sq. ft.

Let x = distance from a. c. to centre of gravity of area to be deducted.

Let Y = distance from a. b.

$$\text{Then } x = \frac{259 \times 7 - 217 \times 7}{42} = 7 \text{ ft. 0 in.}$$

$$\text{And } y = \frac{259 \times 9.25 - 217 \times 9.38}{42} = 8 \text{ ft. 7 in.}$$

which locates the position of the area E.f.g.h., which will give a footing whose centre of gravity coincides with the centre of pressure x.

The footing was then designed for bending by treating it as four beams between the four columns, figuring on 5,000 pounds per square foot upward pressure minus the weight of the concrete in the footing.

While the centre of pressure will, of course, shift under different conditions of column loading, still the variation cannot be sufficient to cause a serious settlement of any part of the footing.

In some cases it is very difficult to economically combine a wall column footing with any other footing. Where this is the case the footing is increased towards the inside of the building and along the wall. When this is done, the column must, of course, be tied in at the top and figured to resist the bending caused by the eccentric loading on the footing. This bending is generally increased by the bending moment from the eccentricity brought on the column from the floor loads.

A point in designing reinforced concrete which is often overlooked is the bending produced in wall columns carrying long span beams. This moment seldom gives trouble in the lower tiers of columns in a building of any considerable height as, in such cases, the columns are so heavily loaded that the eccentricity is not sufficient to produce actual tension in the outside of the column.

The common practice of designing wall columns 20 per cent. heavier than interior columns does not always overcome this tendency to crack from bending, as the extra strength is not applied in the proper place.

Consider the roof columns of a building of considerable width in which the roof beams span from side to side with no intermediate support. The usual custom is to carry the column reinforcement to within a few inches of the top of the roof slab and to bend the anchor bars of the beams down into the columns the usual depth to prevent cracking in the upper surfaces of the beams near the ends. In a building designed in this way the result is pretty sure to be cracks across the outer surfaces of the columns immediately under the level of the bottom of the beams.

even though the roof be under no load other than the dead load of the structure itself. The reason for this is that the beam deflects under its own weight and the weight of slab carried. This deflection produces a tension in the upper surface of the beam at the end, which tension is also present at the outer surface of the column, where it is altogether liable to produce large cracks. These cracks can be seen in many buildings. They should be provided against by increasing the reinforcing steel in the outer side of the column. This reinforcement in these columns (if the same are not bent over as described) increases the liability to crack, owing to the fact that they must be embedded to a greater distance than deformed bars in order to develop their tensile strength. Cracks of this nature are, of course, more unsightly than they are dangerous, for beams supported in this manner are usually designed as non-continuous over the supports and should be of the required strength, whether pinned down to the columns or not. However, the bond with the column is an added strength to the beam and should be preserved.

The placing of brackets under a beam of the above description does not overcome the difficulty and is, in my mind, poor practice. The brackets tend to spread the columns by causing the beam to act as an arch whose thrust is not properly taken care of, and cracks will very likely occur on the outer surface of the columns under the brackets. This construction acts, to a great extent like roof truss without a tie rod.

Reverse bending should be given particular attention in the design of highway bridges where heavy moving loads have to be provided for. In short span culverts, where a flat slab is used, this reverse bending at the abutments, if not properly taken care of, may result in a failure which has all the appearance of a shear failure, and such it may be after a certain point, although it has probably started in tension cracks in the upper surface of the slab.

Consider a culvert, let us say, of 12 feet clear span, to be designed to carry a 15-ton road roller. The slab is designed as non-continuous, and enough steel is inserted in the bottom to give a resisting moment to properly take care of the total maximum bending moment liable to come on the culvert. In all probability the concrete itself will figure to take care of all the shear at 50 pounds per square inch, and therefore no extra provision is made against through shear.

At first glance this culvert would appear to be properly designed to insure against failure from any cause, for, as the slab is not figured continuous over supports, it does not seem necessary to put any steel in the top of the slab at the abutments. This conclusion would be safe if the slab were cast separate from the abutments, but if (as is nearly always the case) the abutments and slab are monolithic, the following is liable to occur:

A heavy, vibratory load comes to the centre of the span and produces considerable deflection and, as the slab is tied down to the abutment, tension is produced in the upper surface of the slab and on the outer surface of the abutments. The slab, being thinner than the abutment, cracks on top just inside the line of the abutment. Then, as the load approaches this point, the shear is increased and the cracked concrete is probably not capable of resisting this shear, and collapses. This failure might have been prevented in three ways, namely, by the use of top steel, by the use of steel shear members, or by having a complete horizontal joint between the slab and abutment.

The advantages of what is known as flat ceiling construction are many, the most desirable among them being the appearance produced and the economy in floor height. The chief disadvantage in the most common types is our lack of scientific data on the subject. In a well-known type, opinions differ nearly 100 per cent. as to the bending moment to be figured in slabs under the same loading. In the Engineering Record of 24th December, 1910, there is an account of some measurements made to obtain the strain existing in different portions of a flat slab floor under working loads. From these strains the existing stresses are figured. The results of these measurements appear to indicate that some designers are over sanguine about the carrying capacities of this type of floor.

A more conservative design of flat ceilings is effected by increasing the width of the beams and decreasing their depth until the under side of the beams is flush with the under side of the slab. The slab in these cases is usually made up of small reinforced concrete joists with tile fillers in between, and two or three inches of concrete over the top to aid in compressive resistance.

In this type of floor the stresses are known and the strength can be figured along the same lines as the ordinary slab and beam construction. The tile fillers are placed as much as possible below the natural axis of the slab so as not to decrease the dead load of the floor. This type of floor is not as economical in steel as the usual slab and beam construction, on account of the decreased arm of the resisting couple of the steel in tension and the concrete in compression.

To-day, reinforcement in a rectangular panel, designed according to the accepted theory of reductions in bending moments, effects economy in concrete only. If the bending moments each way be reduced in the usual manner of multiplying by $\dfrac{B^1}{A^1 + B^1}$ for the shorter span and by $\dfrac{A^1}{A^1 + B^1}$ for the longer, where A represents the shorter span and B the longer, the steel may be slightly reduced by placing less near the edges of the panel than near the centre. This reduction is, however, offset by the fact that, in using bar reinforcement, the amount of resistance of the upper layer of steel is decreased by the decrease in the resisting arm of the forces. The saving in concrete is, of course, effected by figuring it to take its full working compression in two directions at right angles to one another.

Before closing, I would like to enter a plea for the standardization of unit stresses and formulae in reinforced concrete design throughout Canada. Some things, of course, cannot be standardized, but such points as ratio of the moduli of elasticity of steel and concrete, the allowable working compressive strength of concrete, both in bending and in direct compression, the limits of Tee action, etc., might be definitely settled and adhered to by all designers. If it is safe in one city to design a continuous beam uniformly loaded to resist a bending moment of one-twelfth WL, then it is equally safe to do the same in the next city, despite the fact that the second city insists on it being designed for one-eighth WL. Other points might also be straightened out, such as whether a specification should insist on using 12 for the ratio of the modulus of elasticity of steel to that of concrete, when in another part it calls for working stress of 350 pounds per square inch for the concrete in a column and 10,-000 pounds for compressive steel embedded in the same column.

INSPECTION OF CONCRETE.

E. A. JAMES, A.M., CAN. SOC. C.E.

Although concrete was a material of construction in the days when the Roman Empire flourished, yet we speak of it as one of the modern materials of construction. For several centuries other materials were more convenient and more fashionable, and with the settlement of America the plentiful supply of timber and the ease with which it was made suitable for construction work, marked timber the cheap material of construction, and it was not until the '70's that cement and concrete again entered into construction work in any great quantities.

To-day, the annual consumption of cement in Canada is close unto five million barrels, which represents in completed structure many times five million dollars' worth.

Where the uses are so varied, as are those to which concrete is placed, it is but natural to expect to hear of failure, and where the amount of money expended annually is so large, it is reasonable to suppose that every effort should be made to ensure good work.

In concrete work, as in other classes of construction, good design and good materials are necessary, and, granting that we have good design and good materials, a class of well-skilled workmen should produce a finished material of a high order.

Concrete, however, has suffered more than any other building material because of inferior workmen, and this has been brought about largely by those interested in the early days in the sale of cement, who claimed as one of the chief reasons why it should be largely used, was the cheapness of placing, because it did not require skilled laborers.

Fortunately, we do not hear so much of the claim as formerly, for it is well recognized that workmen can, with the same ingredients, produce various grades of concrete.

Although among the first of the modern concrete structures many were well and substantially made, yet there were a number of serious failures, and because the engineers themselves were not any too familiar with concrete construction, and because of this recurring failure and the fact that most of the workers in concrete were men imperfectly skilled, a system of careful inspection was inaugurated.

While the necessity of inspection depends to a certain extent upon the nature of the material and the production, it is always a question of just when and where and to what extent it pays to inspect. In concrete work inspection must be made of the component parts, the sand, the gravel, the cement and the water must be examined, as well as the concrete, while it is being mixed and poured.

The scope and possibilities of a system of inspection enlarges greatly with specialization.

Inspection should be planned to accomplish at least expense, the best results, which may be numerated something as follows:

1.   To prevent loss or defects by accidents or delays.

2.   To prevent loss of time and material on work already beyond repair.

3.   To prevent the necessity of replacing defective work.

4.   To prevent decrease in quality because of the demand for increase in quantity.

5.   To point out imperfections in alignment, methods and material.

6.   To record proper allowances for unavoidable extras.

7.   To draw the attention of the superintendent to workmen who must be better instructed or trained.

8.   To stimulate good-will through fairness, in fixing responsibilities.

Inspection organized to cover any one or all of these purposes will be similar in personality, varying only with the degree of perfection required.

Before it can be determined just when and where it pays to inspect, the following conditions must be satisfied:

1.   Responsibility must be fixed with certainty.

2.   The inspection must not cause unnecessary friction.

3. The inspector must have to do with equality only, not design.

4. The responsibility for defective work must be placed upon the workman as well as upon the inspector.

Responsibility must be fixed with certainity.

As inspection has for its purpose the pointing out of the defeets, it is necessary for the inspector to be able to locate the cause of the defect. One of the most foolish things that can be done is attaching blame to the wrong person, and unless it is possible to discover immediately just when and where the cause of the defect lies, the fixing of responsibility is very difficult. It is, therefore, necessary to have the material on the ground in sufficient time for thorough inspection before it is mixed, for even after the inspector has detected the defect, the responsibility is not necessarily fixed. The error may be due to wrong specification, poor material, defective measurement, defective mixing, or even unsuitable weather conditions. It is, therefore, necessary that instructions and specifications must be in writing.

Inspection must not cause uncessary friction.

No system of inspection which would simply complain of defeets, without attempting to trace the cause, or to assist in the improving of conditions, will be of any assistance. It certainly would not tend to happy relations between contractor and engineer. So it becomes necessary, if full benefits are to be derived from rigid inspection, not only to point out the defects, but the inspector should be in a position to trace the cause and to suggest a remedy. Defective work must be detected as soon as possible, so that the conditions under which the work was done may be fresh in the workman's mind, and the responsibility with certainty attached to him when the defect it through careless workmanship.

The inspector must have to do with quality only, not design.

To point out defects will not necessarily stop the repetition and, although it may be the duty of the inspector to trace the cause, fix the responsibility and suggest the remedy, the inspector must not have to do with applying the remedy or of interfering with the workmen.

When the inspector has reported defects in material or workmanship to the engineer and contractor, or their representatives, he must content himself with awaiting the corrections through the proper officials, although it should be within his power to stop or reject the work until there is an opportunity for investigation, and to take upon himself these responsibilities, he must have knowledge equal to that of the superintendent of the work.

Responsibility for defective work must be placed upon the workman as well as upon the inspector.

. While we have stated that the inspector should not interfere with the workman, yet the knowledge that he is present will have a disciplinarian effect and will prevent the sacrifice of quality for quantity. Inspecting alone will not reduce bad workmanship to a minimum, but the workman must be supplied with proper tools, proper instructions, and must be trained in his work and held responsible for the quality of his work. He must be trained to inspect his own work. We have known cases where the men were paid a bonus for saving cement, and where this is the case it almost requires as many inspectors as men to secure compliance with specifications. Where it is known that the contractor is encouraging his men to skimp the work, the inspector should lay his information before the engineer, and at once vigorous measures should be taken to remove such contractor from the work, for he will not do good work, no matter what the inspection is.

Plan of Inspection.

There are many plans of inspection, any of which may get good results, and all of which may fail in securing good results. Inspection depends more upon the inspector than upon the method.

You may have inspection by central bureaus which retain men who are experts in their line of work, who report first to their bureaus, or to the engineer in charge of the work, as may be arranged.

You may have inspection by your own local staff or foreman, or you may require such guarantee that inspection at the end of one or two-year periods will be all that is necessary. The plan of inspection to be adopted will necessarily depend on the character of work being carried out.

The Duties of the Inspector.

It should be the duty of the inspector to see that all forms are erected on the lines laid down by the engineer, that these forms are stiff and well braced, and that all material and workmanship are in accordance with specifications. He should look after the removal of forms and see that the concrete is not injured in the removal.

Aside from concrete walks, form work is the most difficult to get properly placed, and it is much easier to develop a good inspector out of a good carpenter than out of a good concrete worker.

If the work is to be done at night under artificial light, it will be necessary to increase the staff of inspectors, for concrete that can be detected in the day time, by color, will not show a lack of proper mixing of materials under artificial light. In fact, where high class work is required, or in finishing surfaces, as a rule, it is better not done at night at all.

The cost of inspection is variable, being in some cases as low as one per cent. of the total cost and as high as two and a half per cent. I think it is a usual thing two per cent. should be allowed for inspection, and good inspection is cheap at that price.

ELECTROLYTIC RECTIFIER.

R. H. HOPKINS, B.A., Sc.

Department of Electrical Engineering, University of Toronto.

The general use of alternating current for the transmission of energy for light and power, and the limitations of this form of energy for some purposes, makes necessary some apparatus for the transformation of alternating current into direct current.

There are a number of methods available; the use of the motor generator set, the rotary converter, and the mercury arc rectifier, being some of them. Where the amount of power to be used is considerable, a motor generator set or a rotary converter will be used. When the amount of power to be used is small, and especially where it is used in charging storage batteries, these machines are not as satisfactory. Here, then, is the field for the mercury arc rectifier and the electrolytic rectifier. The smaller the amount of power necessary, and the more the first cost figures, the greater are the odds in favor of the use of the electrolytic rectifier.

The electrolytic rectifier is a chemical cell for the rectifying of alternating current. Its rectifying action is due to a high resistance to flow of current in one direction and not in the other that certain metals have when placed in some electrolytes.

If two plates of aluminium which have been charged (I will mention this later) are placed in an electrolyte such as a solution of ammonium phosphate, sodium phosphate, borax, etc., and connected to a direct current supply, no current will flow: if, however, we connect a plate of iron and one of aluminium to the same supply, the iron being connected to the positive line, the cell will act as a direct short circuit. It is hardly right to look upon the action of the aluminium as regards flow of current when it is an anode as being due to resistance. It is rather a counter electromotive force, which varies with the impressed voltage up to as much as 600 volts per cell. It also varies with the electrolyte. This counter electromotive force is due to a film of aluminium hydroxide that is formed on the aluminium. This film is very thin, being comparable in thickness to the length of a wave of light, and it seems to be formed of two parts, an insoluable shell holding a gaseous body.

If 110 volts be applied to a circuit consisting of two plates of aluminium in a solution of ammonium phosphate, an enormous current flows that rapidly dies to zero, and the plates are said to be charged. On examining the plates it will be noticed that they have a frosty appearance; this is due to the film that

has been formed on them. The statement that the current drops
to zero is not quite true, for if the voltage is direct you will have
a minute leakage current, and if alternating, a leakage current, and superimposed upon it a condenser current of greater
magnitude, but still of small value. Steinmetz says this current
is about .01 ampere per square inch of plate surface, but does not
say for what voltage or frequency. With plates of 10 square inches surfaces in an electrolyte of ammonium phosphate, a current of
.35 amperes was obtained at 60 cycles, 110 volts, and with the
General Electric Company's lighting arrester solution, a current of .15 amperes for plates of 20 square inches on 110 volts,
60 cycles, was obtained.

The connections commonly used for an electrolytic rectifier
are illustrated in Fig. 1. On account of the impedance of the
choke coil, practically no current flows through it, but intermit-

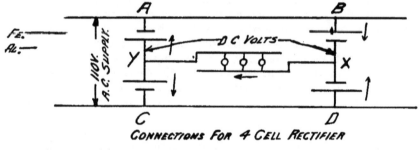

CONNECTIONS FOR 4 CELL RECTIFIER

FIG. 1.

tent direct current flows from the centre of it without a serious
drop on account of the choke coil acting as a one to one series
transformer. Let B be positive. Current cannot enter the solution by an alumium electrode so it flows to centre tap of the
choke coil through the direct current circuit, to the iron plate, to
the aluminium plate on the other side of the line. Now, this current flowing in the half of the transformer creates a magnetic
flux, and unless this flux is balanced the current cannot flow freely. This flux is balanced by the flux created by a secondary or
induced current to the other half of the choke coil, which is
equal and opposite to the first current. It flows from centre point
of coil through the direct current circuit to the iron plate, to the
aluminium plate, and completes its circuit in the coil. Hence
you will have twice the current at half voltage in the direct current circuit. These relations are theoretical and are not obtained in practice.

The connections for the use of an aluminium rectifier without
a choke coil are illustrated in Fig. 2. There are four cells, each
consisting of an iron plate and an aluminium plate in the electrolyte. Two cells are connected in series opposition with the
aluminium plates connected to the supply, and two in series op-

position with the iron plates also connected to the supply. Con_
sider the line AB positive. Current flows from B to X to Y to C,
and if the line CD be positive, current flows from D to X to Y to

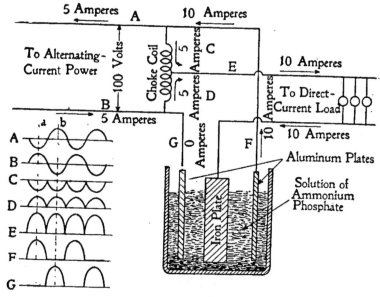

FIG. 2

A. In other words, there is direct current flowing from X to Y
independent of which line is positive.

The results that can be obtained from a small rectifier with
connections made as illustrated in Fig. 1 are shown by a set of
curves, Fig. 3. This rectifier had two aluminium plates two
inches by seven inches, and an iron plate six inches by seven
inches. The electrolyte was a saturated solution of pure ammonium phosphate. Figs. 4 and 5 are taken from oscilligrams
of the voltage and current in different parts of the circuit. The
alternating current supply was from an old, smooth core, ring-
wound alternator, which explains the smooth curves. With a
modern slotted armature alternator as a supply, the general
shape of the curves is similar, but they are somewhat distorted
by the harmonics caused by the teeth of the alternator. These
harmonics are considerably amplified by the condenser action of
the rectifier. Curve 1, Fig. 4, is of the supply voltage; Curve 2,
the supply current, and Curve 3 shows the direct current voltage.
Curve 4, Fig. 5, shows the direct current obtained, and Curves 5
and 6 show the components of 4, i.e., the currents in lines G and F.
(Fig. 1). The theoretical values of the currents in the different
parts of the circuit are illustrated in Fig. 1, the dotted line " a "
indicating the values of the currents in the different parts of the
circuit (which are lettered to correspond) at the instant current
at B is flowing to the choke coil. The dotted line "9" indicates

the values of the currents at the instant current at B is flowing away from the coil.

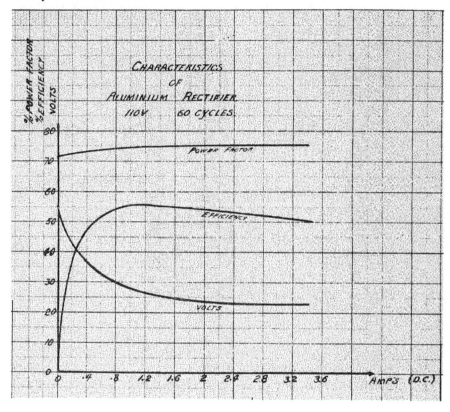

FIG. 3

These rectifiers are used commercially for charging batteries for electric automobiles, batteries for use with pipe organs,

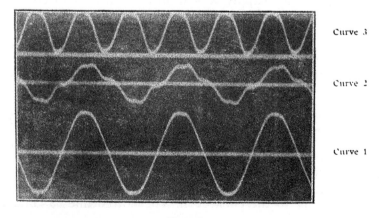

FIG. 4

and batteries for ignition outfits. Their main advantage is low
first cost. The disadvantages are poor efficiency and large size.
The efficiency is about 60 per cent. as a maximum, 50 per cent.
being a good commercial efficiency. As to size, a rectifier for
continuous operation, delivering three amperes, requires alumin-

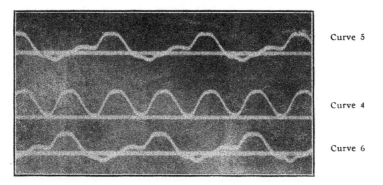

Curve 5

Curve 4

Curve 6

FIG. 5

ium plates of twenty square inches each, and an iron plate of
from seventy-five to one hundred square inches, besides which
the cell must have a radiating surface of from three to four
square feet. This radiating surface, unless some device for cooling is used, is essential, as the efficiency drops rapidly with a
rise in temperature. The power factor is never above 90 per
cent., but is not necessarily low, if the rectifier is fully loaded.
The maximum voltage that can be used per cell is about one hundred and seventy-five volts. Above this voltage the breakdown
voltage of the film is reached. The breakdown voltage may be
increased by a change of the electrolyte, but this lowers the efficiency of the cell.

As a cheap and good means of charging a few storage battery cells from an alternating current supply, the aluminium rectifier seems to give good satisfaction.

SEWAGE EFFLUENTS AND PUTRESCIBILITY.

T. A. DALLYN, B.A. Sc.

Engineer in charge Ontario Board of Health Experimental Station.

The subject under discussion, "Sewage Effluents and Put-
rescibility," is one to which our attention is being constantly at-
tracted as Canada — more especially our Province — becomes
more and more thickly populated.

As our friend and professor—Dr. Ellis—has told us, there
was a time when one might conveniently dispose of all refuse by
throwing it out of the window or around the back entrance. In-

deed, it is only some fifty years since such was the custom in many places in England. In many of the fishing villages the accumulation of trade refuse, such as offal, was such as to make our present-day aesthetic standards stand back in fearsome horror. As was natural in conditions such as these, due to carelessness and lack of all sanitary precautions, disease and epidemics abounded. It was also natural for the sanitarians of that day, when the theory of the spontaneous generation of the organisms of putrefaction held sway, to believe that these rotting masses also produced the cause of the epidemics, the bacterial organ of disease being as yet only dimly suspected by some few workers in surgery.

We find many references, at this date, to the atmospheric conditions during these epidemics, and lengthy assertions upon the effect of the night vapours. Science and observation have since taught us that while these conditions, when prevailing for a considerable period, may lower one's resistance to bacterial invasion, yet they themselves are not the primary cause of the disease. Looking back, one may see many ways in which such conditions may affect the life of a community: (1) Drought lessening the dilutions of stream contamination; (2) Seasonal migrations to the upper reaches of streams which, lower down, act as water supplies; (3) Stagnation of streams, with the subsequent production of algae growths forcing whole communities to seek for other sources of supply where these secondary sources may receive graver contamination than the neglected one. Extent of turbidity and taste are not good nor safe guides to a hygienic water supply.

Pasteur, in his work on yeasts and ferments (1862), discovered a method by which he could sterilize all manner of putrefactive material by heat. Then using vessels that could be sealed while the contained fluid was still sterile, he demonstrated the keeping qualities of the putrescible material. This demonstration was the gravestone of the "theory of spontaneous generation," and the foundation of the latter work in Bacteriology. Like Pasteur, if we are to accomplish much along sanitary lines, we must have results to back up our theories. For example, if Hypo-chlorite is an efficient sterilizer, we must see a reduction in water-borne typhoid infection; in other words, we must look to our death rate reduction; with that proof in hand, we need little else. Now, as a matter of fact, Hypo-chlorite is a good disinfectant. Yet, until it was used in fairly large quantities, .6 parts per million, no appreciable effect was shown on Toronto's typhoid rate. Now, laboratory experiments would indicate that the use of 0.6 parts of available chlorine was unnecessary, and that 0.35 parts would be just as effective; but the practical demonstration showed very different results. Whether this was due to lack of data or insufficient measurements of water consumption, the author is not aware, but the facts are as stated. Hence in carrying on disinfection experiments, one must not place too great confidence upon the laboratory side, except as indicators. The place to look for the effect of your operations is on your sheets showing presence of typhoid, cholera, and other water-carried infections. You have had good reason to know that 0.6 parts available chlorine impart a very objectionable taste to the water. It is this objectionable feature, in connection with disinfection of water supply, that brings us back to our original subject.

If we can not readily sterilize without objectionable tastes a highly polluted drinking water, then we must get after the causes of this pollution and **disinfect** it in such manner that subsequent dilutions will remove all traces of the objectionable tasting disinfectant used. .

Slow sand filters have done. and are doing, splendid work, but they are rather expensive to operate when high degrees of bacterial purification are required. Especially is this true where a lake water, such as that of Lake Ontario, is used, which carries a very fine sand in suspension during storms. The water when in this condition may plug a filter up in three or four hours. Then it becomes a race to keep enough filters clean to supply filtered water to the community. If the water was known to be fairly safe (i.e., all large polluting streams disinfected), filters of different size sand could be adopted which would lessen the frequency of cleaning, and allow the use of various mechanical devices to assist in removing the clogging surface—devices which, under ordinary requirements of filters, are prohibited because they either remove too much sand or else remove too much of the fine sand used in the filter bed.

Of course, we cannot provide for all the various agencies by which pollution occurs, but we can at least get after those which are so apparent as sewage effluents. Gentlemen, you and I are the men that must see to it. If we who possess the knowledge of the evil and the cure, do not agitate and insist on its use, then who will ? We surely cannot leave it to some happy shots of the reporters of our daily papers—for with conditions varying as they do in different places, the remedy must vary also. There is no cure-all, as some of our political newspapers would have us believe. There must be careful investigation, the judicious use of research laboratories and scientific epidemiological field work, not only in one community, but in all. Our municipal Boards of Health are trying to do what they can with inefficient and under-equipped laboratories, but without the backing of the men who ought to know—the doctors, chemists and engineers— how can they go to the electors and councils to ask for the necessary funds ? Gentlemen, you may not care to put your life into sanitary work, but you can, and ought to, at least keep in touch with it and give it your scientific approval when circumstances permit, so that the public will at least know upon whose shoulders the responsibility lies.

In examining the results of the English analysis of sewage and sewage effluents, one is astonished to see the confidence with which systems are advanced upon the results of weekly and monthly analyses. It is only recently in our Canadian Engineer we had published a lengthy recommendation of the De-Clor process. Their experimental plant was in operation for twenty-one weeks, and in that time only twenty-one samples were taken. Does that not strike you as strange ?

The article lays special emphasis on the removal of B. Coli, so that it occurs only in 100 c.c. sample in the effluent, whereas in the prefiltered water it was present in 1-10 c.c. samples (i.e., a removal of 999 in 1,000) the bacterial count only showed a removal of general bacteria 885 in 1,000. Such results as these call for inquiry as to whether the removal of B. Coli is greater than for general water bacteria. The experiments at Mass. Institute of Technology indicate that it is about the same, and, if anything, a trifle lower. I want you to notice this point because it was further illustrated last fall by some rather curious results which came to my notice. The City Hall were making Bacterial analyses every second day, I believe, on tap water to detect B. Coli —sewage contamination. The Experimental Station was making analyses three times each day for purposes in connection with some experimental filters. Now, while the average of the every **two-day set of analyses** was not unlike that of the city for those days, the average of the three analyses a day was altogether different and showed heavy contamination. I am using this only by way of illustration of the point in question, namely, the value of frequency of analysis. It certainly makes a difference in bacterial work, and unless we, as scientifically trained men, know these things and, by knowing them, modify public opinion, a great amount of money is to be wasted in costly and inefficient apparatus. Remember, I am not saying the De-Chor process is inefficient—in fact, I like the idea. I am only saying that the published test was not extensive enough and somewhat misleading.

In sewage disposal there is, as a rule, two outstanding problems. (1) Disposal of trade refuse; (2) Disposal of domestic and factory sewage.

Now, there is no good reason why that trade refuse item should not be fought out with the manufacturers. Trade refuse is industrial waste. Take, for example, an instance in our own city here—that of the Harris Abattoir plant. At one time there was such a nuisance from their operations that the City Fathers, I believe, gave them notice to either remove the nuisance or shut down. They removed the nuisance and made a profit out of the waste. Now we have nothing entering our sewers from this plant save fluids from **which even the grease that comes**

from washing the floors has been extracted. What they have done under splendid management can be done by any other similar plant under equally efficient management.

With many other industrial wastes the problem is as yet unsolved, leaving splendid opportunities for research work under the direction of our faculty and among our graduates. Once the problem of trade wastes in a given centre has been partly solved it becomes a simpler matter to handle and purify the Domestic Sewage.

At present we have many processes for nitrifying sewage. From sewage farms and intermittent filter plants under proper management come splendid statistics, the sludge is digested, the effluent is clear and stable, only requiring disinfection as a final touch.

The question to which I would like to draw your attention particularly is still in the embryonic form as yet, but I bring it rather as a student to students, since it seems to me to be full of possibilities, especially for our Great Lake cities. I think we are generally agreed that in this enlightened age disinfection is a necessary feature of all sewage disposal problems. (1) Because it is not so very expensive; (2) The benefits to be derived are very great indeed in proportion to the expense.

Assuming disinfection necessary, can we not do away with this laborious nitrification, where large bodies of water are available for dilution ? If no process of purification and no disinfection is used, then we get conditions such as we have in our shameful Toronto Bay. This is not sanitary, nor is it a credit to any community in this twentieth century. Suppose we remove the suspended solids present in sewage by some process of sedimentation or chemical precipitation and then disinfect the effluent, will the effluent be stable in dilution ? That is the question. In the present paper I only wish to call to your attention some of the outstanding facts.

(1) We are seeking disinfection, not absolute sterilization.

(2) For stability, such that further putrification, if any, will take place only in aerobic conditions.

Disinfection is limited to the removal of pathogenic organisms. Of course, in process of doing this, we remove some 99.9 per cent. of the saprophitic bacteria, normally found in putrescible solutions, at the same time. Dunbar (1908) claims—and is no doubt right to within certain limits—" that by far the larger percentage of pathogenic micro-organisms are enclosed in gelatinous masses and attached to suspended matter." It, of course, is dependent upon the grade and rate of flow in your sewers whether these conditions exist at the disposal works.—" Hence any process which removes suspended matter removes also a large percentage of the pathogenic bacteria at the same time." " In some experiments on the removal of suspended matter in Hamburg River water, a removal of 30 per cent. of suspended matter gave a corresponding removal of bacteria, and this would be much more so in case of pathogenic bacteria in sewage." Some experiments of our own with Garrison Creek sewage have failed to show such good results. This sewer is laid on a considerable grade and has a velocity of between ten and fifteen feet per second, so that sufficient disturbance is created to largely remove bacteria from any such attachments. However, our experiments have hardly been extensive enough to speak dogmatically as yet. There is no doubt that in chemical precipitation, where the coagulants are of a colloidal nature, some good results should be obtained in removing a considerable percentage of pathogenic bacteria. Generally speaking, pathogenic bacteria may be expected to be more sensitive to disinfection than the saprophitic forms.

One may raise the question of operation of disinfection upon bacteria in spore formation. Fortunately, very few pathogenic bacteria are spore formers, and sewage is not a media in which spore formation is likely to occur if it has not already taken place. There is one, however, that may be mentioned. That is Enteriditas Sporogenes, a bacterium which some investigators claim gives rise to Diarrhoea, especially when present in milk given to young infants. This form is almost always present in sewage, and gives a fermentation reaction similar to B. Coli. The odors, however, are entirely different. This form has been recovered from sterilized sewage—by sterilization I mean disinfection a removal of 99.99 per cent. of bacteria, say a count of 1,000,000 being reduced to 200 and 186—which is a practical possibility) so that we have reason to believe it is more resistant than normal to chlorine. It is an anaerobic bacterium, and this is no doubt why it has been overlooked. If we were to take a known positive sample and inoculate eight fermentation tubes, we might only get four or five positives to show up, due possibly to the fact that anaerobic conditions varied in the several fermentation tubes. However, this is a little aside from the general question, where there is every reason to believe that forms such as B. Typhoid are entirely killed out, and they are the ones to which most attention has been directed.

In some experiments of our own, 1.2 parts per million A. Chlorine served to kill out one million bacteria (suspended in water) so that the water became sterile in 55 minutes. Of course, organic matter has an effect on chlorine usually from eight to ten parts per million are required to practically sterilize Garrison Creek sewage in one hour. These results are borne out also by the work of Earle B. Phelp at Boston some few years ago (1906). I think we may justly assume that the pathogenic bacteria are almost entirely removed, especially where sedimentation or fine screening have been used previous to disinfection. It is worth nothing that H_2S reacts readily with chlorine. H_2S is not as a rule present in raw sewage except as a trade waste from tanneries. But it is very much present in septic effluents—due to the decompositions of the proteids present, no doubt. Hence it is more economical to add chlorine to raw sewage than to septic tank. There are other reasons also why disinfection is preferable before septic treatment, such as the growth of anaerobic bacteria, etc. Septic treatment is, of course, unnecessary where we

Water		Sewage 1-10 dilution
Chlorine	3 to 10	6.5 to 11.5
Nitrates	.43 to 4.95	Raw Sewage .04 and less Nitrified " 0.3 to 3.0
Nitrites	.0135 and less	R. Sewage .01 and less N. Sewage .3
Free Ammonia	.063 to .009	R. Sewage 3.9 to 1.0 N. Sewage .19 to .01
Album Ammonia	.066 to .007	R. Sewage .65 to .01 N. Sewage .01 and less

NOTE.—Values in Table a little high for sewage
and a little low for river water.

are using sedimentation tanks, and has, I think, been generally abandoned in favor of them, where large plants have been installed. Before leaving disinfection, let us remember one thing more, that is, that the chlorine deodorizes the sewage, so that if it be discharged two hours after disinfection we have (1) no odor, or, at least, very little, (2) the bacterial count per cc. varying from 100 to 500.

If you will notice the accompanying table, some must observe that sewage in dilutions of 1-10 shows very little worse than a normal river water. It has, in fact, been offered to some English investigators as pure water. I do not recommend it myself without thorough disinfection, which they overlooked without serious consequences.

Having noticed the constituents of a dilute disinfected sewage, we may turn to its possible stability.

(1) We know that even this dilute sewage will serve to support high bacterial growths, but experiments have only been made at laboratory temperatures, that is, 18° C. and 37° C. Now, the normal for Lake Ontario is about 7.5° C. to 9° C. in summer, except, for surface waters, which may run to 16° C. With them, however, we have storms, and consequently higher dilutions. Then again these subsequent growths must take place in an aerobic media. In this connection, I would like to call your attention to several theories with regard to putrifaction. (1) A known fact is that obnoxious putrifaction takes place only in media where the supply of oxygen is exhausted, when we have such productions as H_2S and NH_3 formed, which otherwise becomes SO_4 radicals and nitrates. (2) These operations only take place due to very high bacterial counts. Alfred Fischer

(1900) states in his work, "The Structure and Functions of Bacteria," the following (see pages 99 and 100):

"Putrefaction is a purely biochemical process, and can only take place when the fundamental conditions for all vital action are fulfilled. If the temperature sinks below a certain point, organic substances cannot putrify, as was well shown by the frozen Siberian mammoths. When discovered, their flesh was so little changed that it was eaten by the hunters' dogs; yet it must have lain in nature's refrigerators for countless centuries. In all methods of preservation the fundamental principle is the same, namely, to create such conditions that bacteria cannot live; for putrefaction—the splitting-up of the nitrogenous constituents of organic matter—is the work of bacteria, and of bacteria alone.

" The list of putrefactive products is far from being complete, for even the qualitative investigation of the processes is still unfinished; quantitative analyses are at present impossible. We do not know, for instance, what determines the predominance of one or the other intermediate product. The effects of the presence of oxygen are somewhat better understood. If air have free access, putrefaction may go on without any odor at all, the evil-smelling gases (NH_3 and H_2S for example) being oxidized at once to form nitrates and sulphates. Aerobic bacteria, too, bring about this mineralization of organic matter, such as the nitre and sulphur bacteria. Moreover, when air is circulating freely there is no accumulation of intermediate products such as skatol or indol. This kind of decomposition proceeding without offensive smell, may be termed decay, as distinguished from putrefaction."

Now, do you see where we have arrived at in our theorizing? If we kill out those forms causing anaerobic putrefaction by disinfection and change of media so that only aerobic decay can take place, have we not arrived at a condition similar to the ordinary respiration of plants and animals, where energy is obtained from the combustion of a few molecules instead of that present in septic tanks, where the energy is derived from the incomplete combustion of many molecules ?

Gentlemen, it seems to me we have a theory here that is worthy of investigation, and I hope at some future date to give you some of the experiments that will be adopted in trying it out, and our results therefrom.

AERIAL NAVIGATION.

CHARLES H. MITCHELL, C.E.

When asked by the President of the Engineering Society of Toronto University to prepare a paper for the Society on the progress of aerial navigation the writer was inclined to seek an excuse in press of other work. It occurred to him, however, that it was due the Society about this time to present a review of the development and progress of aerial navigation, as arising out of his paper on "Aerial Mechanical Flight," read before the same Society in January, 1895, and printed in Vol. 8 of the Society's proceedings. This paper was a resume of the science and art of aerial mechanical flight—if art it was—up to that time, and was compiled from all the sources of information then obtainable, which were scarce, indeed.

The principles enunciated in this early paper are now of especial interest in the light of the intervening history of aerial navigation and of the recent extraordinary successful operations of "heavier than air machines." The chances of the commercial development of mechanical aeroplane flying machines based upon the "everyday principles underlying the kite, the boomerang or the skater on thin ice" were then carefully studied and it was stated that "the perfection of aerial flight will come gradually, as did other perfected inventions which have revolutionized the whole world. We cannot look for any one man to thoroughly solve the problem, but it will be evolved from many sources, and these will at last contribute to the one long desired end." To still further quote this early academic brochure of sixteen years ago, the following significant enumeration of the requisites for a flying machine was made:—

"1. Its various parts and members must be of the lightest construction compatible with strength and stiffness, and the factor of safety must be large.

2. Its general configuration must be economical for space and convenience and present the least possible resistance to the air.

3. It must be capable of rising gently but swiftly and supporting itself in the air in storm or calm for a length of sion."

4. It must have stability and be incapable of upsetting.

5. Should be easily steered in any direction.

6. Provided with a means of rapid and powerful propulsion.

And the sentence is added: "This enumeration may appear highly idealistic, but the practical possibility is much clearer than is generally supposed."

That was sixteen years ago. It is needless to now draw attention to the present-day parallels of these requisites, or to ask to what extent they are fulfilled in to-day's aeroplane prac-tice. But the art is only in its infancy, and who knows what another sixteen years will bring about? This is perhaps even harder to conceive when the progress of building and flying aeroplanes during the past three years is considered in the course of the general evolution.

The purpose of this paper is to briefly outline the develop-ment of aerial navigation, especially in mechanical flight, dur-ing the past few years, and to present some slight description of present-day uses and possibilities—particularly from a prac-tical viewpoint.

Although primarily dealing with flying machines heavier than air, it is interesting to note historically the singular parallel evolution of the dirigible balloon which has almost kept pace with that of the aeroplane.

It is convenient to distinguish between the various classes of air craft according to their nature and functions. The most recent classification is about as follows:—

I. Craft lighter than air: "Aerostatic."
   1. Kites—
       Simple.
       Man-lifting kites.
       Balloon kites.
   2. Balloons—
       Captive.
       Free.
   3. Dirigible balloons—
       Non-rigid types.
       Semi-rigid types.
       Rigid types.

II. Craft heavier than air: "Aerodynamic."
   1. Aeroplanes—
       Monoplanes.
       Bi-planes.
   2. Helicopteres—
       Vertical lift machines.
   3. Combined dirigibles and aeroplanes.

While the main attention herein is given to dirigible balloons and to craft which are heavier than air, a few words may be said in passing with regard to the employment of kites and captive balloons, especially in warfare.

Kites.

Kites have been developed considerably in the British Army and Navy and on account of their cheapness and com-pact form for transportation, have produced very good results in the field. It appears that man-lifting kites have been very successfully operated from ships and lately Major Baden-Powell has worked out a system for using explosive kites against air ships. The man-lifting balloon kite, which is a combination of a small gas-bag and a second bag open freely to the air, has been developed considerably by the Germans, and has been used extensively in their manoeuvres.

Captive Balloons.

The captive balloon is, of course, familiar to all for purposes of general observation, signalling, directing gun fire and, recently, at sea, for the detection of submarine attack. It would appear that notwithstanding the rapid introduction of aeroplanes and dirigibles the captive balloon and kite will still remain a useful means of observation both on land and sea for some time to come.

Dirigible Balloons.

Although many early attempts were made to propel balloons, no real success was made until the recent development of the light, but powerful automobile engine. After types of these engines became established their various applications to aerial navigation was most quickly taken advantage of with the recent remarkable results.

A German named Schwartz is credited with driving the first rigid airship with a gasoline motor, which was 12 h.p. His ship, however, was wrecked after several successful trials. This was in 1897. In 1898 Count Zeppelin came on the scene with a rigid type, having an aluminum frame and gas bags between an inner and outer envelope—built in gas-proof compartments as it were. It was a large vessel 300 feet long, driven with two gasoline motors, each geared to two propellers. Though he secured a still air speed of 16 miles per hour, there were many defects found and the attempt was temporarily given up, but patriotic Germans came forward in these discouraging days with money to build a second.

In 1902 Santos Dumont, in a non-rigid cigar-shaped balloon, performed the first really signal dirigible feat by circling around the Eiffel Tower in Paris, winning a prize of $20,000. This year also saw the first work of the Lebaudy brothers, who brought out a semi-rigid type by which the bending and buckling strains are taken off the gas envelope by a metallic keel; this was propelled by a 35 h.p. motor. This vessel was wrecked in 1903 after doing some 50 successful trips, the longest being 62 miles at 22 miles per hour average. But this was the start of the great French airship fleet, for shortly afterwards the French Army adopted a similar type, and after several successful ships, the "Patrie" was launched in 1906, and in 1907 was used in the manoeuvres, doing a trip from Paris to the frontier.

a distance of 150 miles, at 22 miles per hour. Shortly after.
however, the "Patrie" was wrenched from her moorings, blown
away and lost on the North Atlantic. This was followed by
the "Republique," the largest of the semi-rigid type yet built.
having cylindrical stablizing gas bags at the stem; she was 210
feet long, had an 80 h.p. engine, a range of action of 500 miles
and could carry nine men. In 1908 she did 147 miles at 21 miles
per hour.

In the meantime Count Zeppelin had succeeded in his second ship, which was tested in 1906, but was wrecked by a storm.
This was shortly followed by a third, and in 1908, by a fourth
much more powerful than its predecessors. This "Zeppelin
IV." was built on special specifications of carrying power,

ZEPPELIN IV. DIRIGIBLE BALLOON
Courtesy John Lane Co., from " Airships in Peace and War." *Hearne.*

speed, endurance; she was 446 feet long, had two Mercedes motors of 120 h.p. each, carried a crew of 18 men, and her estimated range of action was 800 miles. One special feature was
the arrangement of the 16 independent gas bags within the
envelope. This airship was tried out in numerous preparatory
short trips before the official government trial, by which the
airship was to carry 16 men and be capable of travelling for 24
hours. In one of these trials in June, 1908, "Zeppelin IV." went
across the Alps, doing in 12 hours a total of 270 miles at an
average speed of 22 miles. In August of the same year an event-
ful attempt in which the ship again travelled a distance of 270
miles at an average of 24 miles per hour, resulted in its com-

plete wreck by explosion and fire; this accident could not have occurred, it is said, had there been apparatus for properly anchoring the vessel to the ground while at rest.

The same year two other successful German dirigible airships, the "Gross" and the "Parseval," both military ships, made their appearance, the former remaining aloft for 13 hours, and reaching an altitude of 4,000 feet, and the latter for 12 hours.

In 1909 dirigible airships of the foregoing types performed many successful and extraordinary flights. France put four notable new ships out—similar to the "Republique." The latter vessel, while going to the French Army manoeuvres, met with an accident which was promptly repaired by the Airship Corps of their army in the field under virtual war conditions. This field repair marked a new step in progress. The ship was used in the manoeuvres and did very useful intelligence work, particularly discovering a wide turning movement of its opponents. The "Republique" was completely wrecked by the breaking of a steel propeller blade in September, and the crew of four killed.

In Germany 1909 saw remarkable performances of the new Zeppelin's, the "Gross" and the "Parseval." The former made a round trip of 800 miles, including the sailing over Berlin, and this placed the Zeppelin far ahead of all rivals. The work of four airships at the German military manoeuvres that year was extensive, and though no information was given out it is known from attaches' reports that exceedingly useful work was accomplished. One interesting operation was a night attack against the fortress of Ehrenbreitstein on the Rhine near Coblenz, in which several ships were employed.

During the year 1909 a new Italian airship in a run of 190 miles made 27 miles per hour average, which captured the high speed record.

The year 1910 produced some new records of particular interest, and there were several notable flights of historic value. Wellman made a courageous attempt to cross the Atlantic, starting near Boston. His arrangement, which he called an "equilibrator," which dragged in the water to stabilize the ship. nearly caused a fatal ending; as it was, the ship was blown about and out of its course by a fierce gale, and was finally abandoned about 200 miles at sea, the crew being taken off by a steamer under thrilling circumstances. The French dirigible. "Clement Bayard II.," made a remarkable flight from Paris to London on October 16th, doing 246 miles at an average of 41 miles per hour, with a crew of seven men. This airship is 251 feet long, 44 feet diameter, has two engines of 120 h.p. each, a range of 750 miles, and carrying a capacity of 20 men.

Dirigibles were used in the principal European Army manoeuvres in 1910, with varying success. In this connection the great success of the British dirigible, the little "Beta," is of interest because of the plucky work of its commander. It is just announced that this airship, working near Aldershot this week, kept in touch by wireless during the whole of a trip of many miles from start to finish.

On February 7th, 1911, the German dirigible, "Gross IV.," was taken out for its first trial. It is 344 feet long and expected to be one of the fastest yet constructed, being capable of making 40 miles per hour. The British admiralty, however, has just now (February 18th) about completed a monster airship—the first aerial "Dreadnought"—at Barrow-in-Furness, 510 feet long, 48 feet diameter, and having 706,000 cubic feet capacity; eight cylinder motors, with three new type propellers are expected to drive the ship at 50 miles per hour.

Aeroplanes.

The perfecting and the employment of aeroplanes is much more recent than the similar progress with dirigible balloons.

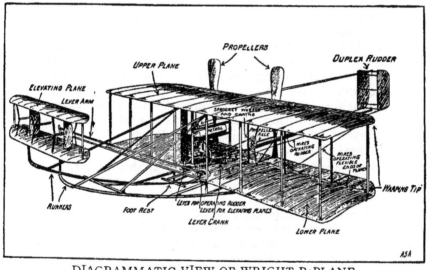

DIAGRAMMATIC VIEW OF WRIGHT BiPLANE
Courtesy John Lane Co., from "Airships in Peace and War," *Hearne.*

The early experimenting and research, however, commenced about 1892, and by 1896 there was considerable data and some experience accumulated with respect to bird-flight, gliding on air and laboratory aero-dynamics. The outstanding features of this period were the experiments with kites by Professor Langley in America, the construction of a steam-driven aero-plane by Sir Hiram Maxim in England, and the actual air-gliding by Lilienthal in Germany. With the latter's death and the great difficulties encountered by Maxim, progress almost ceased, and for a period of eight years the only work done was quiet experimenting in seeking after suitable engines, propellers and forms for aeroplanes. Chief among these workers were the Wright brothers, who, for some years prior to 1904, were working with one and two-plane gilders in North Carolina at a place where among rolling sand dunes a steady wind was assured. In this work they were assisted on the technical side by the late Mr. Octave Chanute, an eminent consulting engineer of Chicago. In such a manner they became expert in the handling of their air craft.

At this period —1905—several forms of aeroplanes had become notable, and with the solution of the engine problem following closely· on the development of the automobile engine.

FARMAN BIPLANE
Courtesy Crosby, Lockwood & Son, from " The Art of Aviation," *Brewer*.

actual flights were accomplished. There remained, however, the perfecting of innumerable details and the gaining of experience and skill on the part of operators to attain the confidence. presence of mind, and almost intuitive quickness necessary to control a heavier than air machine in much the same way as the learning to ride a bicycle. These early experiments of either gliding or driving a plane against the air for a short distance were based upon the principles of soaring bird-flight or of the skater on thin ice.

In 1905 the Wrights astonished the world with the announcement, without details, that their bi-plane machine had actually remained in the air for a half-hour, and later that they

CURTISS BIPLANE USED BY J. A. McCURDY. '07. (TORONTO)

Courtesy The Copp, Clark Co., from "Vehicles of the Air," *Longheen.*

had flown 24 miles in 38 minutes. In other flights they had attained great speed, the greatest having been 38 miles per hour.

As the Wrights undoubtedly led the world in the development and operation of their aeroplane, there are several features of the machine deserving of special mention here. The frame was of hickory and the planes of strong fabric; the wing warping device on the corners of the planes, worked by wires over pulleys for balancing and facilitating turning, were especially novel. This flexure of planes in conjunction with vertical and horizontal rudders enabled the balance to be quickly —almost instinctively—made. By their long experience in these early days the Wrights became so dexterous that they were for some years far ahead of other aviators in their skill in flying. They showed that it was more in the man than in the machine that success lay.

In 1907 a new aviator, Farman, appeared in France, and

BLERIOT MONOPLANE
Courtesy Crosby, Lockwood & Son, from "The Art of Aviation," *Brewer*.

he accomplished numerous short flights, up to a half-mile, in a bi-plane, known then as the "Voisin." His performances, however, were soon eclipsed by those of Delagrange, another Frenchman, who, in 1908, flew various distances up to 15 miles, done on September 6th. But this month of September, 1908, was destined to become notable in aviation, as the Wrights, one in Europe, and one at Fort Meyer, in the United States, were almost daily performing something new, the one breaking the record of the other. The performances comprised flights of over an hour by Orville Wright in America on September 9th and 12th; in the latter 45 miles were covered. Wilbur Wright in Europe on September 21st, flew one hour and a half, in which 56 miles were done and on September 28th, he carried a passenger.

In October of 1908 Wilbur Wright, with a passenger, did 36 miles in 56 minutes, and Bleriot first appeared with his small monoplane, in which he did 3 miles in four minutes and a half.

The year 1909 was notable in aeroplane performances as well as for dirigible balloons. Orville Wright carried a passenger 45 miles in one hour and thirteen minutes on July 22nd, and three days later the world was startled by the news that Bleriot had boldly crossed the English Channel in a small monoplane, 31 miles in 40 minutes. Then on August 26th, Latham. a new aviator, with an Antionette monoplane, flew 97 miles in 2 hours and 13 minutes, and the following day Farman, again to the front, with his bi-plane, broke all records by going 112 miles without a stop. Again on November 3rd, 1909, Farman in his own bi-plane with a Gnome motor, flew 145 miles in 4 hours and 18 minutes. Another significant performance by Farman was on August 28th, when he carried two passengers 6 miles in ten and a half minutes.

This year a second American aviator came prominently before the world; this was Curtiss who, in a bi-plane of his own design performed various feats, especially at the time of the Fulton celebration at New York. His machine, developed with the advice of Dr. Graham Bell, is, in general, similar to the Wrights, but instead of warping the ends of planes, he has small auxiliary planes at the outside ends between the two main ones; his manipulation is also interesting, as he employs the shoulders and swaying body in actuating the rudders for horizontal turning. It is in this type of machine (perfected in the Hammondsport Experiments) that J. A. D. McCurdy, of the class of 1907, Engineering, in Toronto University, is now doing such wonderful feats.

The year 1910 did not produce any extraordinarily long flights, but the altitude records were very much increased, about 7,000 feet being the highest. The various aviation meets, notably those at Belmont Park and Atlantic City, brought out results in control and handling of aeroplanes which prove beyond doubt that these machines are capable of various rapid manoeuvres far beyond the earlier expectations, and these are probably only the beginnings, so that we are justified in expecting wonderful results in stability, manoeuvre and carrying power within the next five years. Speeds were also much increased, especially with the monoplanes which are, of course, the fastest types; Morane, with a new Bleriot, flew 66 miles per hour at Rheims.

As examples of manoeuvre in 1910 two performances in America are notable. One was by Graham White, an English aviator, who flew over the City of Washington, alighted in the

street in front of the Navy Headquarters Building, made a call
and rose again from the street and flew away again over the

GNOME ENGINE AND PROPELLER
(THE WHOLE REVOLVES AS A FLYWHEEL)
Courtesy Crosby, Lockwood & Son, from "The Art of Aviation," *Brewer.*

city. The other was at Belmont Park, when an aviator, off
to a false start on the race course, was recalled and suddenly

circled in a small radius around the judges and the announcement board, back to the track in front of the grand stand.

The flight of Chavez in a Bleriot monoplane over the Alps from Switzerland to Italy in September, 1910, is also notable with respect to manoeuvre, as in 25 miles and a rise of 3,000 feet, he encountered all kinds of vertical cross air currents and bitterly cold air off the snow-clad peaks.

It is likely that the next few years will produce aeroplanes of much greater carrying power as well as of greater manoeu-

INTERIOR OF GNOME ENGINE

Courtesy Crosby, Lockwood & Son, from "The Art of Aviation," *Brewer*.

vring capabilities. Increases of speeds are also to be expected, especially with the monoplanes. It was announced in 1910 that a new racing Bleriot had been built and was being secretly tried, in which speeds up to 75 miles per hour were expected. This type had ingenious wings, which could be flattened out in mid-air and contracted so that the machine could be speeded up while actually in flight.

Combination Types of Airships.

There is not much yet to be said respecting airships combining the features of balloons and aeroplanes. Several such arrangements have been built and tried, but not worked out over long courses. Vertical lifting machines have also been built, but as yet have not become practical. It is likely, however, that considerable progress will be made within the next few years along the combined lines, especially for meeting conditions where ascents and flights are required to be made irre-

GERMAN 3-INCH GUN FOR ATTACKING AIRSHIPS
Courtesy Crosby, Lockwood & Son, from "The Art of Aviation." *Brewer.*

spective of weather conditions, such as may be absolutely necessary in warfare.

There is no doubt that very shortly a vertical lifting "heavier than air" machine will be brought out. capable of standing stationary, or hovering over any point; this will probably combine the horizontal speed properties of the aeroplane.

A new combination of airship and hydro-plane is also beginning to appear, brought about by the necessity of aeroplanes alighting on or rising from water. Only last month Curtiss. in such a machine at San Diego, Cal., alighted on the water alongside a U. S. warship, and after 15 minutes' visit, rose again from the same spot.

Feasibility of Aerial Navigation.

From the foregoing it is not only apparent that the navigation of the air is feasible by dirigibles and aeroplanes, but that as each year passes, with its improvements in types and increased skill in handling, aerial navigation will, before long, be as assured, and as universal as motoring on land or water. It is now only a question of a few years before dirigibles carrying cargoes of many tons, and aeroplanes carrying four or five people will be an established thing. The science will be then beyond the experimental stage and there will be many operators of all nationalities having the requisite experience and skill to actually navigate the air with ease, confidence and safety.

In view of the present state of the building and skill in flying, we are reasonably justified in expecting that in the next two years:—

Dirigible balloons will have a range of action of 1,000 miles, a speed of 40 miles per hour, and a carrying capacity of 4 tons.

Aeroplanes will have a range of action of 200 miles, a speed of 50 miles per hour, and a carrying capacity of 800 pounds.

Both classes of air craft will be capable of operation at will in any moderate wind, either with or against it.

If the foregoing results can be realized the successful permanent employment of such dirigible balloons and aeroplanes for military purposes is an absolute certainty because they are then brought into the category of practical fighting equipment of modern armies and marine navies.

Employment of Airships.

It is as yet premature to attempt any serious conjecture with regard to the ultimate employment of air-craft, either in extent or variety. While commercial uses may appear probable within the next few years military uses are already in sight and a discussion of the possible employment of air-craft for this purpose would be of interest.

Already, as noted, the great powers have been employing dirigibles and aeroplanes in connection with both army and navy manoeuvres. Just now comes the news that Germany will, in the 1911 manoeuvres to be held on the Baltic coast, use flying machines in connection with combined operations in which the battleship fleet will co-operate with their army corps in problems involving the landing of an army in coastal defence.

In those features of modern war, involving tactics and strategy, the employment of air-craft will entirely revolutionize the science. The application of mounted reconnaissance for both tactical and strategical purposes can be applied equally well to aerial scouting with the addition that the range of action and the horizon will be very much increased.

The peculiar adaptation of dirigible balloons and aeroplanes is very marked; a few of these properties, especially for war operations, are as follows:—

1. Wide range of action is now obtainable.

2. Carrying capacity is sufficient for men and food, etc.

3. Speed is as fast as any land or sea travel without delays occasioned by latter.

4. Height of operation is such as to be clear of accurate effective gun fire.

5. Direct routes are available day or night, or in fogs (within limits).

6. Positions of altitude are most adapted for observation and signalling, and for locating submarine objects.

7. Air operations cannot be guarded against except with similar craft or by special terrestrial apparatus.

In adapting these various proven properties of airships and especially of aeroplanes, there are certain well-known uses which have already become apparent for military and general service; an enumeration of these follows:—

For Military Service.

1. Peace and war time reconnaissance, reporting and study of foreign countries, fortifications, harbors, etc.

2. Signalling and wireless telegraph purposes.

3. Carrying despatches.

4. Guards and patrols at frontiers and before an army.

5. Preventing an enemy's observation and screening operations from view.

6. Directing and observing artillery fire.

7. Destroying stores and raiding harbors, fortresses and cities.

8. Surprise or night attacks.

9. Discovering and destroying submarines and mines.

10. In conjunction with general engagements on land or sea.

For Commercial Service.

11. Despatch carrying for emergency purposes.

12. Rapid express transport or mail service.

13. Passenger service of rapid and luxurious character.

14. Exploration in inaccessible or far-distant countries.

15. Scientific research.

16. Rescue purposes at sea.

17. Recreation, sporting and spectacular.

The commercial use of aeroplanes, for instance, while as yet conjectural, affords quite as much variety and opportunity for development in types and their employment as does the military use. It must be remembered, however, that in order to adapt air-craft to commercial use they must meet the powerful competition of the present forms of land travel and the opposition of those commercial organizations operating the land and water transportation utilities. On the other hand, because of the fact that all aerial travel can be made on routes direct and unhampered, and that travel routes have three dimensions in which to operate, the ultimate success of aerial navigation can be fully expected though it may take a stretch of the imagination at present.

As a case in point illustrating commercial use the following comparison in time of the employment of an aeroplane in competition with a railway train, and a motor car, is suggested, it being assumed that a doctor in London is suddenly called upon to make a special emergency call in the country 100 miles distant by bee-line, 120 miles distant by rail, with nearest station four miles, and 140 miles by road. This outline is an extract from "Airships in Peace and in War," by R. P. Hearne:—

SPECIAL TRAIN		SPECIAL MOTOR CAR		SPECIAL AEROPLANE	
Getting ready	30	Getting ready	10	Getting ready	15
Passengers' time to		Journey to doctor's		Passengers' time to	
starting point	15	door	15	starting point	15
Time getting clear		Delay in getting			
of London	10	clear of London	35		
120 miles at 50 miles		130 miles at 35			
per hour	145	miles per hour	222	100 miles at 60	
From station to		Delay in getting to		miles per hour	100
house	20	house	8	Landing and get-	
				ting to house	20
Minutes	220	Minutes	290	Minutes	150

Operation of Air-Craft.

As to the probable methods of operation of the various types of airships much can be said and conjectured, but until a good many features of endurance, reliability, speed and handling are tried out, it is not likely that definite conclusions can be reached. Even with the more stable types of airship, and in the short years of trial up to the present, the accidents which occurred, and the loss of life, have been appalling. In the year 1910 the number of famous aviators who have lost their lives has been most deplorable, but unfortunately it is to be reasonably expected that there will be still many more accidents and loss of life in the strife for the mastery of the air before the art of building ships and flying them will become fixed like other similar operations. As, in the nature of events, the "heavier than air" machine is undoubtedly destined to become the ultimate means of aerial locomotion, it is evident that in its development there must yet be years of trial, success and failure before final definite success is attained.

The various difficulties and dangers which have already been encountered are really at the present time increasing rather than being reduced, for as the art advances and navigators become bolder, the hazards taken are greater. For instance, at one time it was thought that navigation in wind and rain storms, fogs, etc., was impossible; now we find ascents being frequently made under such weather conditions, as, for example, when Latham in 1909 went 75 miles per hour in a gale at Blackpool in his Antoinette monoplane. Fires and explosions on dirigible balloons are a great menace—instance the disaster to Zeppelin IV.— possibly lightning would come also in this category. Breakdown of engine or of propellers or steering gear, etc., in aeroplanes is almost fatal, especially in high flying unless the aviator is successful in righting the machine and gliding to earth without overturning; nearly all fatal aeroplane accidents have been due to this mishap though ehere are several notable examples of the machine being brought down safely,—instance, Curtiss at Atlantic City in 1910. Loss of fuel either by leakage, accident or use is another danger. Collision with buildings, trees or other craft is also to be reckoned with.

Organization and Training of Aerial Corps.

It is not at all surprising that, with all this progress and the swift application of aerial navigation to uses of warfare as the first employment, the nations are seriously organizing and training aerial corps. Next to the development of the machine and equipment, the training of experienced expert aviators and aeronauts is paramount. This is harder than it seems for the means of training are limited, are highly expensive, and often produce discouraging results, as have been experienced the past 3 years. The present year, however, sees all the great powers appropriating large sums in their estimates for this purpose. The German Government has planned very large expenditures, and it is now unofficially announced that eleven German universities will, during the summer of 1911, institute lectures on aeronautics and the mechanical principles underlying the flying machine and its operation. The United States has authorized very considerable expenditures in training, and the news just comes that the National Guard of California has authorized the formation of an aerial corps in connection with the Coast Artillery. It is interesting to notice that the British Army estimates for 1911 include a half-million dollars for new dirigibles and aeroplanes and for the expenses of an aeronautic staff. It is stated in newspaper despatches that the British Army will have five dirigible balloons and five aeroplanes available for use the coming summer.

As to training, especially in aeroplane operation, the work

of Mr. Curtiss at San Francisco for the U. S. Army and Navy is of special note, as indicating how he instructs novices to handle a machine. The first operation after mastering the mechanism of machine and engine is to take short hops or jumps of from 50 to 200 feet, but not higher than 20 feet. Then longer jumps are allowed, and then a low flight, skimming the surface or "grass-cutting" as it is called. After this recruits are allowed to fly and manoeuvre, but always over level ground and close to it. Reports say that practical and athletic officers who are accustomed to motoring and sailing learn very rapidly and safely.

Toronto, March 1st, 1911.

DETAILS OF AEROPLANE TYPES

(As in use in 1909 and 1910.)

CHARACTERISTICS	MONOPLANES		BIPLANES			
	BLERIOT	ANTION ETTE	WRIGHT	FARMAN	VOISIN	CURTISS
AEROPLANES:—						
Span, feet	28	46	40	33	38	29
Area, sq.ft.	150	377	540	430	540	250
WEIGHT:—(No Pilot)						
Total, pounds.	462	1045	880	990	1100	550
Per sq.ft. of Plane	3.08	2.77	1.63	2.30	2.04	2.20
MOTOR:—						
Type—Cylinders	3	8	4	4	8	8
Revs. per min.	1200	1100	1500	1300	1200	1200
Power in H.P.	24	50	30	50	50	30
Sq.ft. Area per H.P.	6.25	7.50	18.0	8.6	10.8	8.3
Weight per H.P. per sq. ft. Area	0.13	00.5	0.05	0.04	0.04	0.07
PROPELLER:—						
No. of blades.	2	2	2 of 2	2	2	2
Material.	wood	steel	wood	wood	steel	steel
Diameter	6ft.9in.	7ft.0in.	8ft.0in.	8ft.6in.	8ft.6in.	6ft.0in.
Speed.	1200	1100	450	1300	1200	1200
SPEED:—						
Miles per hour in still air:						
Average.	40	38	39	41	37	48

APPLIED SCIENCE

INCORPORATED WITH

Transactions of the University of Toronto Engineering Society

DEVOTED TO THE INTERESTS OF ENGINEERING, ARCHITECTURE
AND APPLIED CHEMISTRY AT THE UNIVERSITY OF TORONTO.

Published monthly during the College year by the University of Toronto Engineering Society

BOARD OF EDITORS

H. IRWIN, B.A.Sc. . . . . Editor-in-Chief
M. H. MURPHY, '11 . . . . . . Civil and Arch. Sec.
F. H. DOWNING, '11 . . . . . Elec. and Mech. Sec.
E. E. FREELAND, '11 . . . . . Mining and Chem. Sec.
A. D. CAMPBELL, '10 . . . . . Ex. Officio U.T.E.S.
R. L. DOBBIN, '10 . . . . . . Ex. Officio U.T.E.S.

ASSOCIATE EDITORS

H. E. T HAULTAIN, C.E. . . . . Mining
H. W. PRICE, B.A.Sc. . . . . . . Mech. and Elec.
C. R. YOUNG, B.A.Sc. Civil
SAUL DUSHMAN, M.A. . . . . Chemistry
Treasurer: M. B. WATSON, '10.

SUBSCRIPTION RATES

Per year, in advance $1 00
Single copies 20
Advertising rates on application.

Address all communications:
APPLIED SCIENCE,
Engineering Bldg., University of Toronto,
Toronto, Ont.

EDITORIAL

During the last week in April some seven hundred and eighty-five men, undergraduates of this Faculty, their Spring examinations over, will be ready for a Summer's vacation. The many methods of spending it that are in mind about this time of year do not in all cases coincide with

VACATION AND VACATION WORK. the opinions as to how it might be spent to best advantage. Professionally, a man can do no better than begin, on May 1st, and abandon about September 25th, a system of labor which will yield him the greatest possible grounding and advancement in his chosen field of engineering. The advisability of such procedure, even in the most healthful, out-of-door branches, might not pass criticism, and a shorter term might be recommended, and to them who will find themselves engaged in foundry or machine shop, practice in underground or chemical work, it is obvious that an intermission between the clash with the April cavalcade of technical disturbances, and the setting out to vacation work, is also in the interests of young engineers.

Graduation in mining requires that the students shall have at least six months' practical experience in mining, metallurgy, or geology, "for which they must receive regular wages." In the departments of mechanical engineering and electrical engineering, the minimum is set at eight months. These are minimum quantities, and, naturally, the maximum is undefined, the inference being that the requisition is based on the stable assumption that as much practical work as possible should be worked into the university course in engineering. This doesn't include a spattering book canvass of a municipality, or a Summer's cruise on a passenger steamer as waiter. These occupations are acknowledged means of remuneration to students of various other faculties. But if the engineering student has a primary view to reimburse his saving bank, he will best do so by engaging in, and adhering to, the work with which he hopes to fill out a certificate form, whether his department calls for it or not. In other words, the monetary value of vacation work in engineering is, generally speaking, much in advance of any other to which the average student is obliged to turn his hand. Moreover, the advantage of experience in the proper field is obvious.

The undergraduate in civil engineering learns early in his course that experience in the field is a very necessary adjunct if he hopes to derive full benefit. A clause in the calendar requiring a certificate of experience is evidently unnecessary. The more practical experience acquired early in the course, the more valuable the course is to him, and the higher will be the salary that the profession will find him capable of earning upon graduation, is his conception of the problem.

Then again, there are other reasons why the college graduate is not prepared for his place in the world upon leaving college, unless he is already equipped with knowledge gained through practice. For example, the most efficient graduate is one who, during his four years at college, has had a good taste of the work before him when he leaves, not altogether because of the knowledge he has gained by applying technical formulae to that work, but because of what he has learned of life as lived among engineers, and of the problems of discipline that have confronted him. The more he obtains of it, the greater will be his directive force, and the more effective will he prove himself, by his early submission to ordinary, every-day engineering discipline. This is as necessary to the man who is going to be a successful engineer, as financial means is to the man who choses to spend the coming vacation at Newport or Rockaway.

THE ELECTRICAL CLUB.

Though of comparatively recent origin, the Club has, since its inception, held an important place among the student organizations. A few words regarding its history and aims may not be out of place.

It was founded during the term 1906-07, primarily for the purpose of encouraging public speaking among engineering students, by the presentation and discussion of technical papers. Its membership was limited to third and fourth year Mechanical and Electrical students, and the fact that it has flourished in spite of its limited membership is an indication that it is meeting a need of the students. The first president was W. MacLachlan. Since then the presidents have been: 1907-08, F. R. Ewart; 1908-09, C. L. Gulley; 1909-10, C. J. Porter; 1910-11, W. P. Dobson.

It was first known as the S. P. S. Electrical Club, but in 1909 the name was changed to The University of Toronto Electrical Club. Regular meetings are held every two weeks, at which technical papers are read and discussed. It was thought when the club was formed that the members would enter the discussions more freely in a small meeting than in a large one. This has proved to be the case, and to this fact is due in a great measure the success of the meetings.

Among the subjects discussed during the present term were Wireless Telegraphy, Siemen Bros.' Electric Railway Equipment, Automobile Motors, Commercial Testing of Transformers, Electrolytic Electrifiers, The Oscillograph, Multiple Unit Control, Pennsylvania Electrification, Gas Engines.

An important feature of the work of the Club is the collection of the publications of the leading maunfacturing companies. These are placed on file, and form an increasingly valuable source of information to the members.

The idea of holding excursions to points of engineering in the city originated with the Club. These have always been well attended and have served the useful purpose of giving the students an insight into the practical side of engineering.

The executive for the session 1911-12 is as follows: President, C. De Guerre; Secretary-Treasurer, R. Taylor; 4th Year Councillor, R. A, Storey. The Vice-President and 3rd Year Councillor will be elected next Fall by the 3rd Year.

Under these able officials we bespeak for the Club a prosperous year and a continuance of its prestige among the students.

IMPORTANT MEETING.

The 260th meeting of the American Institute of Electrical Engineers will be held in Toronto on April 7th in the Chemistry and Mining Building of the University of Toronto. The speaker will be Mr. W. S. Murray, electrical engineer, of the New York, New Haven and Hartford Railway, who will present a paper entitled "Analysis of Electrification, and Its Practical Application to Trunk Lines for Freight and Passenger Operation." As Mr. Murray is an authority on this subject, the meeting will be one of greatest interest to men in this branch of the profession. The meeting is open to all who may be interested.

THE ENGINEERING SOCIETY ELECTIONS.

P. G. CHERRY, '11

Nominations for office on the executive of the Engineering Society were held on Wednesday afternoon, March 8th, at 3.30 o'clock, in the second year drafting room at the rear of Convocation Hall. It is to be regretted that Convocation Hall was not available for such an important assembly. The Society executive were at a disadvantage by not being advised of the change until almost the last moment, and it can be assured that next year will witness better accommodation for the annual meeting.

Many of the same old "gags" to pull the freshmen votes were sprung as "planks" in the "platforms." But this year has created a precedent in that the President-elect pronounced against the production of problems which had been dealt with, and would be dealt with, merely for the sake of having a platform, for he was elected practically on a non-platform ticket, believing, as he stated, that there would be enough for the next year's executors to handle with the present questions in hand and those which are bound to come up next year, and which cannot be predicted at present.

The two days following the nominations were devoted by the candidates and their supporters to canvassing, and justice was done to the time-honored customs, the presidential candidates, of whom there were no less than four, being exempt from the customs, by unwritten law. Perhaps the posters this year exceeded any hitherto attempted. There were signs of all designs and sizes, some reaching half way across a room, and others draping from floor to ceiling. One candidate enlarged his photo, and used the copies in the various rooms. The introduction of "The Toike Oike," the campaign paper containing principally the advertisements of the candidates and their platforms, etc., was an important feature, being published for three successive days, giving the photos of the presidential candidates in the issue of election day. The idea, we believe, emanated from Mr. J. A. Stiles ('07), B.A.Sc. Mr. R. W. Moffatt ('05), B.A.Sc., was the editor-in-chief, and Messrs. L. S. O'Dell ('07), B.A.Sc., and L. T. Rutledge ('09), B.A.Sc., were the associate editors.

As in previous years, the candidates were given the opportunity to speak on the day of elections in drafting room " A," the freshmen quarters. The elections took place on Friday, March 10th, the polling being held for one hour in the afternoon at the Engineering Building, proceeding in the evening at the gymnasium from seven to eleven o'clock.

That Friday evening at the gym. typified the school spirit beyond peradventure. It can no doubt be said that the enormity of the occasion has increased in proportion to the increase in the undergraduate body at the school, until the efficiency of the function approaches very nearly 100 per cent. A graduate of many years, who has not looked in for some time, cannot realize the development.

A mud bath was one of the most prominent features. With

W. B. McPHERSON

a full force of " water-boys " operating between the gym. and the heating excavations near by, it took fully three hours to satiate the appetite for mud and water. And such a mob the men did compose, many being saturated with the gluey liquid from top to toe! And how proud they all were of their appearance, just as Diogenes was, before them! Voting has always been a privilege, but never was the opportunity to vote at such a premium. The writer had occasion to relate in " The Varsity " of one rising young engineer of an investigative temperament, who discovered a water tank under the roof. Desiring to demonstrate first-hand

UNIVERSITY OF TORONTO ENGINEERING SOCIETY EXECUTIVE, 1910-11

C. V. Perry.　　F. E. Freeland,　　F. H. Downing,　　E. V. Chambers,　　M. H. Murphy,
　　　　　　　　　Vice Pres.　　　　Vice Pres.　　　　　　　　　　　Vice Pres.

I. Irwin, B.A.Sc.,　W. T. Curtis,　R. L. Dolbbin,　A. D. Campbell,　M. B. Watson,　A. H. Munro,　J. McNiven.
　　Editor.　　　　　　　　　First Vice Pres.　　President.　　　Treasurer.　　Cor. Secretary.

H. V. Clark, Rec Secretary.　　　　　F. R. Gray.

Applied Science

INCORPORATED WITH

TRANSACTIONS OF THE UNIVERSITY OF TORONTO ENGINEERING SOCIETY

| Old Series Vol. 23 | APRIL, 1911 | New Series Vol. IV. No. 6 |

IMPRESSIONS OF ENGINEERING IN GREAT BRITAIN.

CHESTER B. HAMILTON, JR., B. A. Sc.

The Institution of Mechanical Engineers, whose headquarters are in London, England, last year invited the American Society of Mechanical Engineers to hold a joint meeting with them, during the last week of July, 1910.

The official party of the American society, numbering about 150, left New York on the White Star liner "Celtic," on July 16th, while about as many more went earlier.

We arrived in Liverpool on the evening of July 24th, and were met at the mouth of the Mersey by a tender, which brought out a deputation from the Lord Mayor of Liverpool, and from the Institution of Mechanical Engineers.

From that point on the time was so crowded with engineering and social opportunities that it made one wish that there were at least forty-eight hours in the day. Our English friends certainly are the kindest of hosts. The convention lasted a week, first at Birmingham, then at London, stopping over a day on the way at different points of interest. All the numerous side excursions throughout the week were arranged so that people of the most varied tastes might choose according to their preferences, and thus everyone might be satisfied.

The convention closed Sunday evening, with a visit to Westminster Abbey, around which we were conducted by the sub-Dean. The engineers were particularly interested in the memorial window to Sir Benjamin Baker, the builder of the Forth Bridge and the Assouan Dam.

The following day the party began to break up, the members going where the greatest interest drew them. The writer went to Holland and Belgium and the Brussels Exposition; then returned to England, visiting as many engineering and manufacturing plants as possible, and finally took a short trip to Scotland.

In general it may be said that in all matters touching on either the navy or merchant marine, one may look for the highest efficiency in British works or factories.

The new Yarrow shops on the Clyde, building torpedo boats

and destroyers, are an example. They are not very large, compared to some of the other great works, but they are as good shops as I have ever seen—high, well lighted and ventilated, good crane service, and general orderly arrangement. One of the features is a roofed-over "water-dock" (this name being used in distinction to a dry-dock), equipped with electric traveling cranes, under which the destroyers are fitted up after launching. Only turbine engines are now used, along with the famous Yarrow watertube boilers. It was rather a surprise to see them building destroyers for Holland—a maritime nation itself. They were also at work on boats for Denmark and Brazil. These latter are taken to Brazil by their own black and mulatto crews, but the Yarrow's send one or more engineers from their shops to insure their arrival, which otherwise might be doubtful.

At Cathcart, a suburb of Glasgow, I visited the Wier Pump Works. They build only direct acting steam pumps of very high grade, principally for marine use. This is another first class shop, with both its general methods and details more carefully worked out than usual. The principal part of the shop is one storey with saw-tooth roof. A line shaft down each row of columns drives the machines in that bay; power is supplied by a horizontal single cylinder gas engine at the end of each bay, using producer gas. Thus they get most of the advantages of group drive, as the number of machines on each shaft is small; and also get very cheap power and high efficiency, for the transmission is short and direct. There are almost no motor drives in this shop. They have little to gain, and the management fears the expense for maintenance of 1,500 small motors. They use a large number of Reeves' variable speed friction gears, where an electrically driven shop would use interpole motors, and by fine adjustment of the speed of the machines are able to get the highest efficiency from the tools. They like the Reeves gear very much, and are applying it to most of their new tools.

A large number of their machine tools are made especially for the job for which they are intended, thus they are simpler, more rigid and accurate, and sometimes cheaper than standard tools, as they do away with unnecessary adjustments, slides, and universal features.

They have a very good foundry for admiralty bronze, where they do first class work. They are under the difficulty of having to use very poor, almost gravelly moulding sand. They have a good system of standardized flasks and large patterns, divided into sections the same length as the depth of one section of the flask. In this way the mould can be given to a number of different men to finish at the same time, thus making greater speed possible. Every convenience is supplied the men to enable them to turn out high grade work efficiently. This foundry was a great contrast to one in Birmingham, making builders' hardware by most antiquated methods. I account for this by the fact that the Birmingham firm were not doing marine work. The Wier Pump Co.'s methods evidently pay, for they claim, with apparent right, to be the busiest shop in Great Britain, employ 2,000 to 2,500 men under good working conditions, and are putting up additions to their shop, while some other people are struggling to keep out of the receiver's hands.

The Vickers Maxim Co. and the Sir Wm. Armstrong, Whitworth Co. are two of the largest iron-working firms in the world. They both make almost everything, but, of course, specialize in work that is too big for other firms.

One of our party was talking to Mr. Matthews, the managing director of the Armstrong, Whitworth Company, without knowing who he was, and asked him what line he manufactured. He replied: "Oh, anything you like," adding in explanation,

Fig. 1   Brick Arch Railway Bridge on the Thames.

"anything from a ten-penny nail to a fully equipped battleship."

In the Whitworth shops I saw a very large and heavy engine lathe, specially built for another company, to machine the parts of large marine steam turbines. This lathe was large enough to swing a small room over the ways, and so heavy that the ground in the purchaser's shop would not support it, and shipment had been delayed till another site could be found. Another machine tool not seen in this country is the armor plate shaper. These machines somewhat resemble a radial drill (on much larger scale, of course), having a very long arm with an auxiliary support at its end. The tool carriage slides on the horizontal ways on the arm (like the drill head), and is driven by a multiple thread lead screw of rather long pitch. A double tool is used, so that a cut is taken on both forward and return strokes. The machine is used for cutting armor plate and finishing the edges. Such great works as these have facilities for setting up in the shop the complete side armor of a battleship, to insure its accuracy of fit before it is sent to the shipyard.

In the Whitworth shops there are almost no automatic machines. They use a large number of boys on cheap lathes on work that could probably be better done on a screw machine. They do this in order to train up a sufficient supply of good lathe hands. This question of an extensive apprenticeship system versus automatic tools seems to be one of the weak points of the American method, which tends to produce a shortage of really skilled mechanics.

At the Vickers Maxim shops I noted the remarkable way in

Fig. 2. Crossing Gates on an English Railway.

which government work is done in an open shop and yet kept secret. I saw many parts of H. M. S. "Lion" and another ship of the super-Dreadnought type, and was told that the information that such a ship was being built had only become known after practically all her parts had been completed.

At Wier's Pump Works, at Glasgow, I was shown the condenser pumps of a great ship, which they were forbidden by the purchaser to speak of except by its shop-order number. The jealousy between rival trans-Atlantic steamship companies assists in the maintenance of secrecy, for no one could tell whether these pumps were for a Cunard or White Star ocean greyhound or a British or a Brazilian Dreadnought, though it was evident to anyone that they belonged to a record-breaking engine-room equipment.

It is extremely difficult to keep any secrets now that Japanese inspectors on Japanese Government work are in the shops, for they go everywhere with sharp eyes and notebooks. Vickers used to build complete ships for the Japanese. They made the guns for their latest Dreadnoughts, but the Japanese are building the ships themselves. Soon the Jap will be making guns and all.

A rather interesting machine tool is in use at Vickers (and probably elsewhere) for cutting the portholes in armor plate and for boring out large ingots and forgings preparatory to hollow-forging them on a mandril for the parts of heavy guns. Its ac-

Fig. 3.   Entrance to the Cathedral, Antwerp.   Fig. 4.   Motor Busses, Trafalgar Square, London.

tion is quite analagous, but on a much larger scale, to a diamond rock drill. The boring bar is hollow and carries a series of cutters arranged around the end so as to cut out an annular channel and leave a large core, thus effecting a large saving of time and power.

These great shops, Vickers, Maxim's and Armstrong, Whitworth's, are too big for appreciation or comparison by any standards we have in Canada.

Reference was made above to a foundry in Birmingham,

where very poor methods were in use, both mechanically and in the treatment of the employees. A considerable proportion of the latter were women and children, who were working under "sweat-shop" conditions. This place, I understand, was typical of many others. Piece work was the rule, and the rate was so low that they had to work in desperate haste for long hours to earn very small wages.

Malleable iron castings were being made in small crucibles (60 to 100 lbs.) The moulders worked in groups of three under a contract system in little half-open sheds around the side of the yard. No moulding machines were used. None of the patterns were carded, all being in loose pieces, and the moulder had to cut the gates for every little piece, weighing, perhaps, only a few

Fig. 5. Willans & Robinson's Engineering Works, Rugby.

ounces, every time it was moulded. In the electro-plating room there was no efficient ventilation. The operators, who were young girls, were working in a choking atmosphere. What little machining there was on this class of work was done by child and female labor, in a half-lighted basement, with poor tools and, as always, at extreme speed.

In going through the factory districts there was opportunity to observe much more of the poverty of the country than is seen by the ordinary tourist. And it is not the poverty of hard times or accidental misfortune, but a deep, hopeless state, from which there seems no escape. What encouragement can these "submerged" ones take from the news that the Ontario manufacturer, the Manitoba wheat farmer, and the new Transcontinental railways are in need of labor? None, for that is on the other side of the world, and, besides, many of them are mentally and physically unfitted by heredity and environment.

One impression of the whole country is that it is a land where people and human effort and human suffering is cheap and material is expensive, in contra-distinction to America, where raw material is cheap and people are relatively more important. In Holland and Belgium dogs are much used for light draught. One American engineer said: "This is a fine country, but I'd hate to be a dog here." So I would say that England is a great and wonderful country, but I'd hate to be a laborer or

Fig. 6. English Goods Truck.

mechanic there. I thought many times of Goldsmith's lines in "The Deserted Village":

"But a bold peasantry, their country's pride,
When once destroyed, can never be supplied."

Messrs. Webley and Scott occupy, for the manufacture of revolvers and automatic pistols, a rather old, crowded, and badly arranged building, and the speed of production is high, but the quality of their output is of the highest, and the workpeople seem intelligent and comfortable. Is the explanation of this that theirs is the trade of the armorer, which, with its companion, the ship-builder, holds such a premier position in Britain? In this large shop there was not a drawing or a scale to be seen, outside the tool room, everything being made to standard jigs and gauges.

The illustration of the Willans and Robinson shop, at Rugby, shows the handsome building and the fine gardens and shrubbery about it. The works are arranged so they can be enlarged to double the present size by extending sideways without disturbing any of the departments. All this reserve is given to the workman for garden plots. A large and elaborate testing plant has been provided, adjoining the power house, and connected to the rest of the works by a three-foot gauge railway system. Here careful tests can be made on engines, turbines, and pumps under almost service conditions, before shipment. The departments in the remainder of the works are conveniently arranged in series, as follows, from rear to front: Pattern storage, foundry, casting storage, forge and machine shops, erecting bays, packing and forwarding department, pattern shop, and office. It will be noted that, with the

Fig. 7. The Dock Cranes at Antwerp.

exception of the pattern shop and office, this is the logical order in which work would pass through the shops.

The famous Willans engines were formerly made here in large numbers, but of late the principal output of this company has been steam turbines, Diesel engines, turbine pumps, and condenser apparatus.

Traveling so much away from the beaten track of the tourists gave a good opportunity to compare European with American methods of handling passengers. To be quite frank, they were generally not such as to excite admiration. We have all heard how the pirates used to make their victims "walk the plank." A modern repetition of this may be seen in the method of unloading second and third-class passengers at Queenstown, which resembles also the way cattle are handled at the large packing houses. Rain, wind, and the hour of 3 a.m., all added to the situation. The "victims" were compelled to carry their own hand-baggage, which averaged more than two pieces per person, down a long, narrow, slippery gang-plank, poorly provided with cleats and rails, at a slope of between 40 deg. and 45 deg. One end of the plank swung with the motion of the ship, and the other with that of the tender, which was to take the passengers ashore.

English trains accelerate more rapidly than our heavy American rolling stock. In view of this it was a real disappointment one day to find that a train which was already half-way out of the station could be overtaken and boarded. The fact that a fine or imprisonment is provided for boarding a train in motion was not known then.

In the eyes of an American the compartment system is alto-

Fig. 8. Working on the Dykes, Holland.

gether objectionable. The best trains now are constructed on the corridor plan, but they are so narrow that they are not in the same class as American coaches. The complete absence of any baggage checking system is a very bad feature of English travel. The responsibility rests entirely on the passenger to see that his baggage arrives at the same place and time as himself. In this connection it was necessary to use the long distance telephone several times, and I found both the service and the clearness of transmission far below what would be tolerated in American or Canadian practice. Telephones are not used in England to anything even approaching the extent they are in this country.

English express trains are equipped with the Vacuum brake, but many passenger cars and all freight cars are without brakes. The illustration shows an arrangement, which is called a brake, used only for holding a freight car on a siding. Some freight locomotives have only hand brakes. Automatic couplers are unknown, the three chain links being universally used. On passenger equipment the middle link screws up like a turnbuckle, drawing the cars together and compressing the spring buffers. This makes the train very smooth in starting. In coupling cars an implement called a coupling hook is used which resembles a garden hoe with the wide blade cut off and the iron spike twisted like three-quarters of a turn of a corkscrew. Thus a man need not go between the cars—provided he has a hook with him.

One advantage, however, of the great amount of slack in a

Fig. 9. Dutch Windmill.

train of such cars is that less tractive effort is required to start from rest a train of these than one of equal tonnage of close-coupled cars. Consequently English locomotives are light, particularly as to weight on the drivers, compared with American engines. It seems that this, along with the light weight of the cars is the real reason for the smooth roadbeds of the English railways. Great pains are taken in building the roadbeds, which certainly reduces the maintenance cost, but it is doubtful whether they would stand the extreme hard usage to which American roadbeds are subjected much better than ours do.

The oddity of shipping merchandise, often perishable, in open "goods trucks," covered only with tarpaulins, strikes the Canadian visitor strongly. These "goods trucks" and "goods vans" are exactly like the pictures in an old book I have which was published over thirty years ago. Apparently there has been no substantial change or improvement in this rolling stock in the last thirty or forty years. The ordinary goods truck is ten or twelve feet long over all, and carries ten or twelve long tons. To ship a sixteen-foot board requires two cars, unless it can be arranged to overhang the next load. The small size of the standard tunnels is one of the great obstacles to the English railways.

After all, no matter how far the English railway falls short of American and Canadian standards and requirements, we must remember that it fills the English requirements satisfactorily, and carries the inland traffic of a great nation with greater speed and safety than is the case with American or Canadian railways.

Figure 2 shows an arrangement of crossing gates, which affords great safety, but could probably not be applied in a country having much snow.

American locomotives have been tried in England, but have been found unsatisfactory. The material and workmanship are both much inferior to those in English engines. English engines all have solid copper fireboxes, and many have copper tubes. The English idea seems to be to make machinery as good as possible, whether it is economically worth while or not, while the American way inclines toward building only to last till the time when the design will probably be out of date. Each method suits the conditions of the respective countries.

In urban traffic the horse is rapidly disappearing. The taxi-cab replaces the hansom and the motor-bus its horse-drawn predecessor. The type of motor-bus used in London is shown in the photograph of Trafalgar Square.

Figure 1 shows a brick arch railway bridge of very good design. This compares favorably with some of the recent reinforced concrete arches. The centre span of this bridge is said to be the longest brick arch in Great Britain.

It was said that the only important features in Antwerp were the cathedral and the method of handling freight on the docks with gantry cranes. Whether this is true or not I cannot say, but these two items at least are worthy of the admiration of the architect and the transportation engineer.

Figure 8 shows a rather inefficient but very industrious application of hand labor to the making of additional land on the Zuider Zee, in Holland. The material is dredged up from the bottom of the sea, some sixteen feet, and filled in between the old dyke and a new one (not shown), thus making one more farm.

There is, on a trip like this, an almost unlimited opportunity for American engineers to observe new and useful methods, and we were all deeply grateful to the Institution of Mechanical Engineers for their very great kindness.

BUOYS FOR CURRENT READINGS.

E. R. GRAY, '13

Under instructions from Mr. C. H. Rust, C.E., City Engineer for Toronto, a series of observations were conducted during the summer of 1909, dealing with current movements in Lake Ontario, near Toronto. At the outset it was found rather difficult to obtain a type of float which would give a satisfactory indication of the true direction, and rate of flow of the subsurface current without it being unduly affected by wind or surface motion.

In the first instance a metal float of the type and dimensions shown in Fig. 1 was used.

It consisted of a double cone-shaped, air-tight float, made of tin, to the top of which a short length (2 feet) of 1-4 gas piping was attached. This rod carried a small movable sheet-iron flag, by which the buoy was located when in use.

At the lower end of the float more piping was fastened, the length of which depended upon the depth of current to be gauged.

A sheet of galvanized iron 12 in. x 4 in. was then fixed rigidly to the lower end of this rod, to take the pressure of the current.

It will readily be seen that the large area which the body of this float offered to the wind and surface current, was out of all proportion to the small vein attached to the rod below, and that if, in turning, the vein offered its thin end to the direction of the current, there was practically no subsurface resistance plane.

This buoy was soon discarded, as no reliable results of the movement of subsurface currents could be obtained.

The buoy as shown in Fig. 2 was then tried with fairly satisfactory results, but proved so heavy and cumbersome that it soon gave way to a better type, viz., that shown in Fig. 3. The second style of buoy consisted of a small block of oak 4" x 4" x 6", to the centre of which a small swivel was stapled. Varying lengths of copper wire, No. 18, dependent upon the depth for which the buoy was to be used, were fastened to the swivel.

To the loose end of the wire a slight strip of wood, about two feet in length was attached, and to the other end of the strip a small piece of white cotton was tacked, in order to facilitate the location of the buoys on the water.

About the middle of this strip, two or three ordinary net corks were wired in order to carry the strip upright, when the oak block was weighted carefully to the proper depth.

On account of the weight of the oak block, which frequently

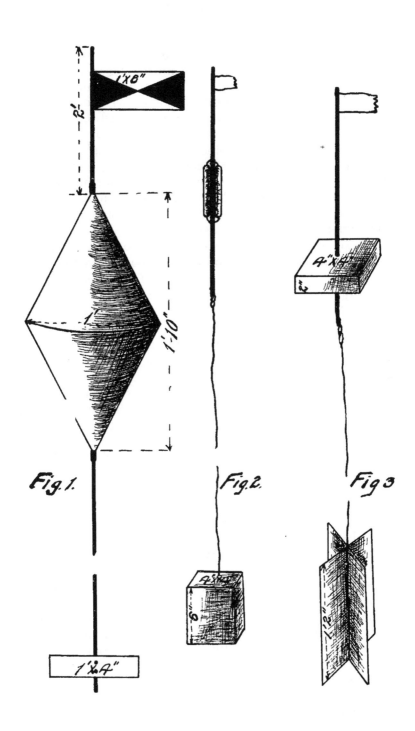

Fig.1.   *Fig.2.*   *Fig 3*

broke loose, this buoy was abandoned, and the type shown in Fig. 3 substituted with better results.

Two pieces of tin 10" x 14" were riveted together down the centre with three copper rivets. The opposite sides of both pieces of tin were then bent through ninety degrees, thus forming two vertical planes at right angles. A hole was punched at the centre of one end and through it a copper wire loop was bent. To the loop was fastened an end of the required length of copper wire, forming the submerged part of the buoy.

The float was made of a block of cedar 4" x 4" x 2", painted to keep out the water. Experiments with other woods demonstrated the fact that cedar gave the best results for buoyancy and economy. In the centre of the block a small hole was bored, through which a strip of wood about 2 feet long was passed, allowing some six inches to project from one side.

To the long end a small piece of white cotton was tacked to act as a flag and to the lower end a small loop was wired.

To complete the buoy the loose end of the wire attached to the submerged vein is fastened through this loop.

It was found that the use of a loop in attaching the wire to the float was a good plan, otherwise the continual swinging and bending motion given to the float by the movement of the waves broke the rigid wire fastening, resulting in the loss of buoys and a negative observation.

With this buoy, then, it will readily be seen that extremely accurate results may be obtained regarding the movement of subsurface currents. The small cedar float and mast offer very little surface to surface currents and wind action, while of the large surface on the lower or submerged part of the buoy, half may at any time offer resistance to the current motion, no matter in what direction the vein may turn.

Larger sizes were tried, but in this instance the results were not shown to be proportional to the addition of metal used.

For observation, a number of buoys of varying lengths were placed in the water together, at a point the exact location of which was known.

They were then allowed to drift, and were located later and collected when deemed necessary. In this case the referencing was done by the use of the sextant. Two angles were read from the buoy upon three objects—one of which was common to both angles—of known position; for instance, certain prominent towers or chimneys. These angles were then plotted on a plan and the buoys located.

This may be done when the buoys do not pass beyond the range of the shore, with good results. It also proved helpful in conducting these observations, to note the speed and direction of the wind, in relation to the movement of the buoys. These com-parisons gave very interesting results.

The observations were carried on under the direction of Mr. F. W. Thorold, B.A.Sc., Assistant City Engineer.

Record of Winds, 1908, Obtained From the Meterological Observatory.

	June.	July.	Aug.	Dec.	Sept.	Oct.	Nov
N.	2	0	2	0	1	0	3
S.	12	1	0	0	0	0	0
E.	15	11	10	14	12	6	2
W.	1	19	16	13	19	24	19
Variable			2	2	3		1

1909.

	Jan.	Feb.	Mar.	April	May.	June.	July.	Aug.	Sept.
N.	7	3	11	4	1	0	1	1	1
S.	2	0	1	2	2	0	0	0	0
E.	4	5	6	11	12	10	10	12	11
W.	18	18	11	8	11	16	18	15	13
Variable		2	2	3	5	4	2	3	3

Of 485 days observed 50.5% were westerly.
" " " " 31.1% " easterly.
" " " " 7.6% " northerly.
" " " " 6.7% " variable.
" 4.1% " southerly.

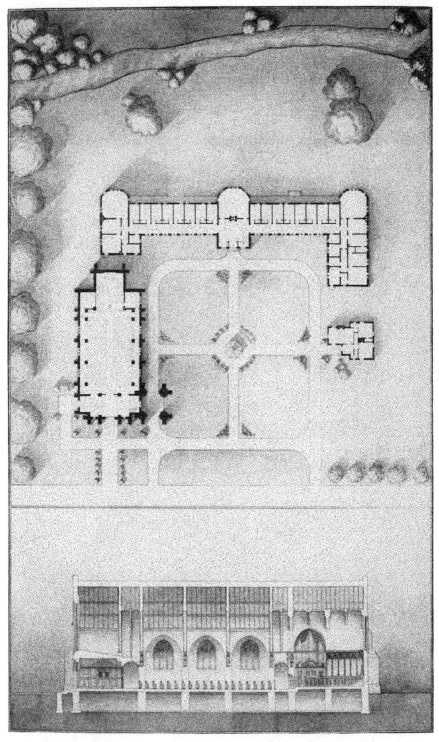

A HOME OF REST FOR AGED POOR, WITH DETACHED
CHAPEL AND MANSE
Designed by T. L. Rowe, '11

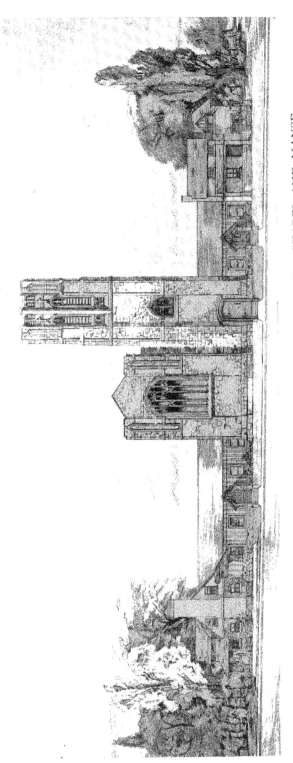

A HOME OF REST FOR THE AGED POOR, WITH DETACHED CHAPEL AND MANSE

Designed by Paul Sheard, '11

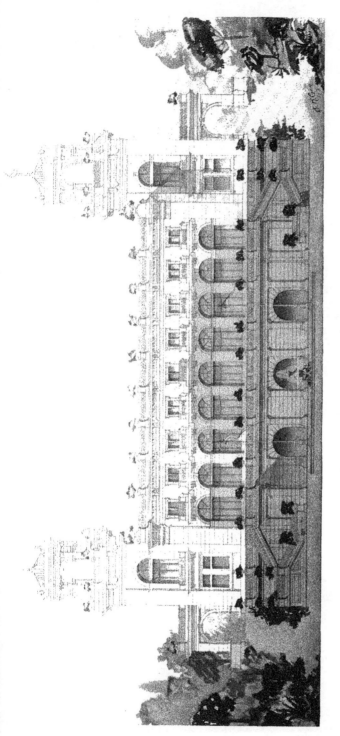

A RESTAURANT IN A PUBLIC PARK

Designed by B. R. Coon.

(Second Year Student)

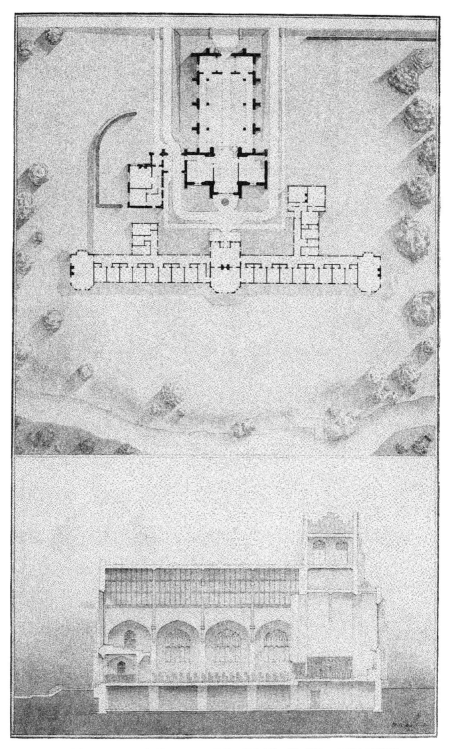

A HOME OF REST FOR AGED POOR, WITH DETACHED
CHAPEL AND MANSE
Designed by H. H. Madill, '11

A SHELTER FOR A FOUNTAIN IN A PUBLIC PARK
Designed by R. S. McConnell.   (Second Year Student)

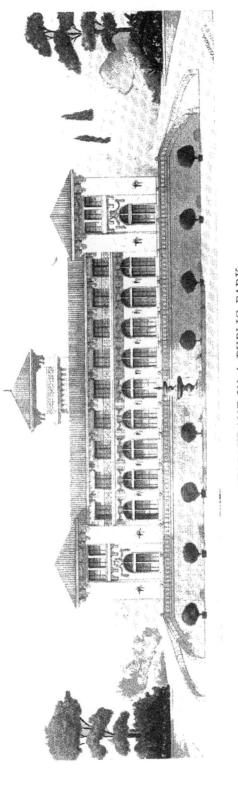

A RESTAURANT IN A PUBLIC PARK

(Second Year Student)

Designed by L. C. M. Baldwin.

A GATEWAY TO A RICH MAN'S ESTATE

Designed by A. C. Wilson.

(First Year Student)

PENNSYLVANIA ELECTRIFICATION.

H. A. COOCH, B. A. Sc.

The problem of transportation in the city of New York is one which has puzzled the minds of the most talented and experienced engineers since 1871.

New York, of course, is so situated that either bridging or tunneling was necessary for a terminal station in the heart of the city.

In 1884 a proposition was discussed to build the "North River Bridge," with a span almost twice that of the Brooklyn Bridge, but due to the enormous cost both in property and construction, as well as the possible obstruction to navigation, it was abandoned.

The tunnel proposition under the North and East rivers

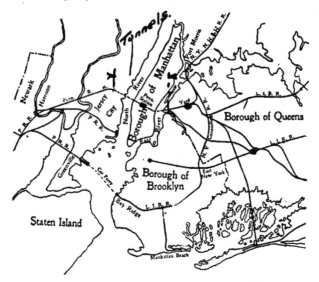

Fig. 1. Sketch of New York and Vicinity.

was often considered, but was out of the question as far as steam locomotives were concerned.

The successful development of the electric motor for traction purposes, first in quite small units, as applied to car axles, and during the past ten years in sizes so large, and capacity so great, as not only to equal, but in the case of the "Pennsylvania Locomotive," to exceed the power of any steam locomotive heretofore built, has solved the above problem.

The electrification is entirely of direct current, which, for a comparatively short mileage, compares very favorably with the single phase.

The length of track electrified is twenty miles, extending from Harrison, N. J., to Long Island City, about twelve miles being tunneled in all. This can be plainly seen by referring to Figure 1.

It has been estimated that the ratio between the coal burned for operating passenger trains by electric, and that for steam locomotives is 1 to 2, which seems to more than counterbalance the necessary initial outlay, while, as far as convenience and cleanliness of operation is concerned, electrical equipment is much more satisfactory.

Thus the Pennsylvania Railroad has fitted up one of the most elaborately and magnificently equipped terminal stations in the world. (Fig. 2.)

Altogether the operations around New York city have cost the railroad nearly $150,000,000.

The new station itself is located on 7th Avenue, between 33rd and 35th Streets, and is handy to all parts of the city.

At present the passenger traffic through the new station

Fig. 2.   Pennsylvania R. R. Terminal Station, New York City.

numbers about 500 trains per day, while the construction has been so made as to allow of a possible 1,000 per day. All the traffic, of course, with a few exceptions, which is handled by the new station, is passenger, the freight being taken across the river by car ferries from Greenville to Bay Ridge.

The locomotive itself (Fig. 3) is designed in two half units. The design of the articulation between them is such that the leading half serves as a leading truck, and the other half as a trailing truck for travel in either direction.

This type of locomotive has also a very peculiar method of drive, which distinguishes it from all others. Instead of the usual methods of drive, i.e., of gear wheels, as in ordinary street cars; quill drive, as in the case of the New York, New Haven and Hartford; or direct drive, as in the case of the New York Central type, connecting rods on the drivers are connected by cranks to the crank or eccentric on the motor, like ordinary steam locomotives. This arrangement can be quite clearly seen in Fig. 4.

The drivers are 72 in. in diameter, and the pony trucks 36 in. in diamteer. The total wheel base is 23 ft. 1 in. per half.

The weight of the locomotive completely equipped is 332,000 lbs., while the weight on the drivers is 207,800 lbs.

The motors (Fig. 5.), two of which are placed on each

Fig. 3.   Sketch of Pennsylvania Locomotive.

engine, are connected for series-parallel operation, but either motor can, without the aid of the other, run the engine should the occasion arise. Each unit is equipped with the Westinghouse electro-pneumatic system of multiple control, so that in a case of a particularly heavy load, as many locomotives as necessary can be used, each controlled by the master controller in the front cab of the leading engine.

The location of the motors is high above the driving axles,

Fig. 4.   General View of Running Gear.

which means that high speeds are possible on turning without any extreme stresses on the rails.

With one motor alone the locomotive can exert a tractive effort of 30,000 pounds, the maximum possible tractive effort being 69,300 pounds, although 79,200 pounds have been developed.

The maximum capacity of the locomotive with the two motors is 4,000 h.p. for short periods, and 2,000 h.p. continuously.

The motors are designed to operate on 600 volts d.c., and each weighs 45,000 lbs.

The maximum speed obtainable is 95 miles per hour, and the maximum guaranteed speed is 80 miles per hour.

The motors themselves carry 2,900 amps., at 600 volts full

Fig. 5.   Detail View of Motor.

Fig. 6. Power Station, Pennsylvania Electrification.

load, and operate at an efficiency at this load of 92.1 per cent. They take their power from a protected third rail, and are also connected to a pantagraph which, for safety purposes, is used together with an overhead construction in the case of railroad crossings too wide for the momentum of the train to carry them across at a reasonable speed.

The motors are provided with commutating poles, consequently have sparkless commutation, even when carrying the momentary heavy currents used in acceleration, which are far in excess of the normal running current of the motors.

Notwithstanding the good commutation resulting from the presence of auxiliary poles, provision is made to relieve the driving gear from the dangerous shocks that would result in the event of an accidental flash-over at the brushes. A flash-over practically stops the motor armature for an instant, and unless the shaft can slip in the armature core the effect on the driving gear would be disastrous.

This is prevented by a slip clutch between the armature core and its spider. The clutch can be set by means of the springs, so that slipping will start at any overload. In the case of most of the locomotives these clutches are set to slip at 100 p.c. overload, and consequently this renders the pins and connecting rods quite safe.

The power for operation of the trains is supplied from the power plant (Fig. 6), located at Long Island City. At present it has a capacity of 40,000 k.w. and is capable of an extension to 75,000 k.w.

The total number of locomotives used by the Pennsylvania is 24, the cost being about $50,000 apiece. These, together with 280 multiple unit cars, take care of all the present traffic.

OBITUARY.

Roy B. Ross.

The death, in far-away Buenos Ayres of Roy B. Ross, '05, a few months ago, lessened our number by one of the most promising young engineers of the Faculty of Applied Science has produced. He received his early education at Ingersoll, Ontario, and afterwards came to Toronto to reside, attending Toronto Junction High School. Mr. Ross entered upon his course in Engineering in 1900, and after completing two years became identified with the Otis Elevator Company of Yonkers, N.Y. In 1904 he resumed his course.

Roy B. Ross

After graduating with honors he became associated with Haney & Miller, in this city. Leaving their employ, he joined the International Marine Signal Co., in their Ottawa office. His engineering ability and ingenuity gained for him a position with the same firm in New York City a short time later. Then in 1909, Mr. Ross went for them to further their interests in Argentine Republic, where he remained until the spring of 1910. He spent the summer in his native city, Toronto, returning to Buenos Ayres, early last fall. It was in the south that his remarkable progress was checked and his early death, at the age of 26 years, occurred.

Mr. Ross was known and admired by every classmate for his congenial disposition and exemplary companionship, and his devotion to the ideals that characterize the successful engineer. In the crystallization of present into past, the personality of Roy B. Ross will long be remembered as a brilliant college man, and a man of promise. For those in Toronto, to whom the sad news of his death in strange and distant Argentine, came as a bolt from blue skies, we extend the sympathy of the class '05 and of the School.

H. Stanley Fierheller, B.A.Sc.

The Faculty of Applied Science lost one of the brightest of her younger members, and the School one of her most promising graduates in the early summer of 1910, when H. Stanley Fierheller passed away.

Stan. received his early training in Markham public schools and collegiate, and after obtaining the highest standing there became one of the most popular members of the class of '05·

Immediately after taking the degree of B.A.Sc., with honors,

H. Stanley Fierheller, B.A., Sc.

he joined the electrical department, taking charge of the lectures in electric circuits to the first year and the laboratory work of the fourth year. At the end of two years he obtained leave of absence and spent a year at research work, publishing at the end of the year an extended paper based upon his researches. He returned to the School in the fall of 1909, but ill-health forced him to resign.

His work and his personality, as shown in the short time he was with us, will be a constant memorial to a real man.

Milton Thomas Culbert, B.A.Sc.

The death of Milton T. Culbert, on March 14th, removed from the engineering profession one of the cleverest men that the Canadian mining industry has ever known. Mr. Culbert was born in Granton, near London, Ontario, on May 29th, 1880. He received his public school training partly in Granton and finished in London, Ontario, where he passed his high school entrance in 1893, at the age of 13. He

Milton Thomas Culbert, B.A., Sc.

spent three years at the London Collegiate Institute, and left school to drift out to the foothills of the Rockies. The following summer he spent in Parry Sound, and while here visited the Calumet and Hecla, and the Bruce Mines, and took a deep interest in mining. Some of his mining friends persuaded him to take a mining course, to which his father readily consented. He passed his matriculation examination in 1898, and spent the next four years in the mining course of the School of Practical Science, graduating and taking his degree of B.A.Sc., both with honors.

His first summer vacation was spent at Copper Cliff; the other three with Prof. Coleman on the geological survey of the Sudbury district. He also prepared the report on the Hutton Iron Range. In 1903-4 he was demonstrator in mineralogy and geology in the University of Toronto. In 1904 he compiled the map of the Sudbury Nickel Range. The following winter he went to the Western States for a short time, and spent some time in the refinery of the smelter at Trail, B.C. Hearing of the Cobalt find he came east again, and met Mr. M. J. O'Brien in the spring of 1905, and entering his service, proceeded to Cobalt, where he opened up and developed the O'Brien mine.

Mr. Culbert had always taken an active interest in the progress of the Faculty of Applied Science and Engineering. Since his taking up residence in the north country he has been at work on a collection of valuable ore specimens, which would have soon been complete, for presentation to the University. He was actively associated, also, in all that concerned the progress of the Cobalt and Temiskaming districts, being a member of the municipal council; a member of the council of the Cobalt branch of the Canadian Mining Institute; president of the Temiskaming Mine Managers' Association; and president of the Temiskaming S. P. S. Graduates' Club, as well as being prominent in Masonic and social circles.

His rapid advancement was checked by a fatal attack of appendicitis, and cut off death at the early age of 30 years.

The funeral in London was conducted under Masonic auspices, and was attended by a large delegation of mining and municipal men from the Cobalt district, showing the esteem and honor in which Mr. Culbert was held.

To Mrs. Culbert and her son, Milton, jr., and to Mr. Victor Culbert, B.A.Sc., '07, brother of the deceased, our heartfelt sympathy is extended.

PRESIDENT CAMPBELL'S RETIRING ADDRESS.

Gentlemen :—

This is the last time that I shall have the opportunity of addressing you as your president. I wish to thank each one of you for your hearty co-operation throughout my term of office.

The past year has seen the coming into effect of some important changes in the affairs of the Society, and with your permission I will review the work of the year.

The Society have had this year for the first time a secretary employed exclusively for its work. Mr. Irwin, who has occupied this position, has performed his duties as editor-in-chief of "Applied Science," and as manager of the supply department, in a manner highly satisfactory to the executive and to the members of the Society.

The supply department has had a record year. A permanent sales clerk has been employed there throughout the year. The new system of book-keeping has been used in this department as in the others. The policy here has been to extend the stock handled and to make only such profit as is necessary to meet the expenses of the Society. We would like again to draw the attention of the staff to the fact that should they desire to publish text books, covering any of the branches of our studies, the Society is now in the position financially to aid in the publication and distribution of such books.

Our monthly, "Applied Science," as you have heard from the editor's report, has had a good year. It continues to grow and promote more cordial relations between graduates and undergraduates, and we look forward to its being issued monthly throughout the year, rather than for but six months of the year, as at present. The monthly has done much to build up the feeling of "Boost for the School."

Our Annual Dinner was a success. At it we entertained officers and members of the Toronto Board of Trade, the chairman and some members of the Royal Commission on Technical Education, as well as representatives from the various educational institutions and technical societies and journals of the city.

A new social function was started this year by the Society. We held the first annual Science dance. The dance was a great success socially and financially, and the thanks of the Society are due to the committee in charge of it.

The papers read before the Society have been of interest and the meetings well attended. Speakers from Chicago, Montreal, and Ottawa, at the expense of considerable time and money, have come to Toronto to address us and to give us the benefit of their experiences. Good papers have been the rule, too, in the sectional meetings, and excursions arranged by the vice-presidents, assisted by the staff, have been attended with profit.

The most cordial relations have existed between the students and the staff, whom we have found always willing to aid in the advancement of the Society. Your thanks are especially due to Dean Galbraith, Prof. Wright, and to our representatives on the University Senate, Mr. E. A. James and Mr. C. H. Mitchell, for their aid in carrying out the work of the year.

One grievance we have, in the cramped condition of our reading room and library. Extension there would indeed be appreciated by our members. Some new engineering books have, however, been added to the central library and sub-libraries this year, and with the adoption of some system of indexed cataloging to make the books more available, we may have hopes for a solution of this trouble in the not too distant future.

I have had reason to congratulate myself on the very able executive with whom I have been associated in carrying on the affairs of our Society. To them belongs the credit for anything that has been accomplished this year. I commend to you for your approbation the members of that executive. I congratulate you on your choice for the coming year, and feel confident that under their guidance the Society's best interests will be advanced.

In conclusion, let me thank you for the honor you have done me, an honor which it has always been my honest endeavor to merit.

I take great pleasure in introducing to you your president-elect for 1910-11, Mr. A. W. McPherson.

THE TREASURER'S REPORT.

Toronto, April 1st, 1911.

Gentlemen—

I beg to submit my report as treasurer of the Engineering Society during the year 1910-11. The system of bookkeeping recommended by the auditor a year ago, and described in the revised edition of the Constitution, has been very satisfactory enabling us at all times to have our financial affairs under easy supervision.

Through the medium of the Supply Department the Society ventured this year upon the publication of a number of booklets bearing upon courses of lectures. Among these were "Notes on Descriptive Geometry for the Third Year," by Mr. J. R. Cockburn; "Notes on Dynamics of Rotation," by Prof. W. J. Loudon; "Notes on the Chemistry of Fuels," by Dr. Ellis, and "Notes on the Calculus," by Mr. S. Beatty, the latter being a small textbook of some seventy pages. The department also began the sale of photographic supplies, and, in short, broadened out extensively to meet the needs of undergraduates.

Applied Science has, during the year, almost succeeded in refunding to the Society the $275 paid to the former editor last summer, this amount having been due him upon commission basis, and having been left over by the former executive. This has evidently been an exceedingly successful year for the monthly.

A summary of the business of the year appears herewith:

Cash Book Balance.

From April 1st, 1910, to April 1st, 1911.

RECEIPTS.

From 1909-10 Executive:

Cash in Bank	$ 324.84
Cash in Till	28.50
Deficit of Cash	2.29
Merchandise Sales	7,563.68
Applied Science	1,377.72
Fees	748.00
Annual Dinner	509.65
Annual Dance	40.75
Telephones	49.75
Outstanding Cheques, March 31st, 1911	440.08
	$11,085.26

DISBURSEMENTS.

To 1911-12 Executive—

Cash in Bank	$ 344.09

Forward ....$ 344.09
Cash in Till 417.49
Applied Science 2,011.24
Supply Department—
    Merchandise 5,513.31
    Salaries 966.33
Meetings and Entertainment......... 229.55
Annual Dinner 715.06
Sundry Expenses 700.44
Telephones 187.75

                                                                          $11,085.26

Balance Sheet.

RESOURCES.

Merchandise, as per inventory Mar. 31, 1911. $1,676.05
Cash on Hand 417.94
Cash in Bank 344.09
Accounts Due Supply Department 70.66
Accounts Due Applied Science 1,078.98
Fees Outstanding 70.00
Dinner Deposits 40.00
Office Equipment less 10 p.c. depreciation ... 212.85

                                                                          $3,910.57

LIABILITIES.

Accounts Outstanding Applied Science 2.52
Accounts Outstanding Supply Department... 91.11

                                                         93.63
Surplus 3,816.94

                                                          $3,910.57

Surplus March 31st, 1911 .... $3,816.94
Surplus March 31st, 1910 .... 2,862.48

Net Gain for Year .... $ 954.46

Respectfully submitted,

M. B. WATSON,

Treasurer.

APPLIED SCIENCE

INCORPORATED WITH

Transactions of the University of Toronto Engineering Society

DEVOTED TO THE INTERESTS OF ENGINEERING, ARCHITECTURE
AND APPLIED CHEMISTRY AT THE UNIVERSITY OF TORONTO.

Published monthly during the College year by the University of Toronto Engineering Society

BOARD OF EDITORS

H. IRWIN, B.A.Sc. . . . . Editor-in-Chief
M. H. MURPHY, '11 . . . . . . Civil and Arch. Sec.
F. H. DOWNING, '11 . . . . . . Elec. and Mech. Sec.
E. E. FREELAND, '11 . . . . . . Mining and Chem. Sec.
A. D. CAMPBELL, '10 . . . . . Ex. Officio U.T.E.S.
R. L. DOBBIN, '10 . . . . . . Ex. Officio U.T.E.S.

ASSOCIATE EDITORS

H. E. T HAULTAIN, C.E. . . . . Mining
H. W. PRICE, B.A.Sc. . . . . . . Mech. and Elec.
C. R. YOUNG, B.A.Sc. Civil
SAUL DUSHMAN, M.A. . . . . Chemistry
Treasurer: M. B. WATSON, '10

SUBSCRIPTION RATES

Per year, in advance $1 00
Single copies 20
Advertising rates on application.

Address all communications:
APPLIED SCIENCE,
Engineering Bldg., University of Toronto.
Toronto, Ont.

EDITORIAL

The April issue of "Applied Science" appears again as a re-
minder to the Alumni that the term is at its close. The undergraduate readers beg to be excused from being
Volume IV. included within the range of such a warning.
Number 6. They need it not; with them the remaining
hours are numbered. As to the trial balance
that the month of April is striking for them at present, and likewise, to the fulness of the coming vacation "Applied Science"
concurs with them in their anticipations and hopes.

In Volume IV. there appears to the editor a goodly number
of deficiencies in the make-up of the Journal. Many of them, six
months ago, seemed so easy to dodge. Like the undergraduate,
he has "hopes," and these are that the shortcomings have not
all bared themselves under the eye of publicity, and that the
publication, as a whole, is not the standard by which the Alumni

measure the particular branch of the University of Toronto in
which they are most interested.

The Faculty of Applied Science and Engineering has, like
the monthly, shown but little outward evidence of material progress over last year. No new buildings have
The   been added. The old red brick, under the name
Faculty.   of Engineering Building, still **adorns** the
southern border of the campus. In fact, to the
daily passer-by, everything seems the same, except for the quak-
ing upheaval that is slowly slicing the grounds to the north and
south, with a view to producing heat and light from the direction of the ravine.

But within doors the year has been one of the most progressive. Each of the seven courses of instruction has made good
progress in establishing better working foundation and broader
scope, and in every department there appears to be at present
the desired combination of the old fundamental principles with
and by methods that are the most modern and best established.
We do not know whether or not this was the policy laid down
a year ago, but, at all events, it has made itself strongly known.

When the copy is closed, this number, enlarged to nearly
twice its usual size, has the proportions which it is to be hoped
it will attain within the next few years. We
List of   trust that upon opening it and glancing through,
Graduates.   the reader will not find disappointment in the
cause of the increase in the number of pages.
The contained list of graduates is taken directly from the Calendar for 1911-12, that is now on the press. Its publication here
will give the members of the classes of '05 and '95 and '85 an idea
of the expansion and of the result of the early and unceasing
efforts of the men to whom this expansion can readily be traced.
It will give classmates the addresses of their classmates, and will
give all to understand that "back home" their names and present
professional capacities are on record. We ask each one to aid us
in keeping this record as up-to-date as possible. This only requires the mailing of a postcard to inform us of any change of
address. Professional news of the graduates is also very much
appreciated.

The appearance in this issue of further examples of work
done in the Department of Architecture need not be explained.
The illustrations speak for themselves. That
Archi-   this course in the University of Toronto has
tectural.   been making rapid advancement is the word left
by leading architects who have visited it re-
cently. These men being qualified to judge by the correct standards of the present, their approval bears considerable weight.

The few examples of design shown here are supplementary
to those which appeared in the February issue, and are indica-
tive of the work done in the four years of the course.

There are approximately two dozen of our graduates in Montreal. The number increases yearly, and, during the summer months, is nearly twice as large, owing to the presence in that city of many of our under- graduates. Just at present there is some little talk of an S. P. S. Club being formed. The eu- terprise would undoubtedly succeed. The clubs in New York and Pittsburg are extremely well founded and well recognized. The S. P. S. Temiskaming Club is an energetic organization, that does not hesitate at inadequacies of location or environment. It has been an important pillar in the progress of the Cobalt region, and the credit is due to the enthusiasm and efforts of the late Mr. Culbert and his colleagues.

S. P. S. Club in Montreal.

In Montreal there are men of the old "School," who have kept in touch with us as well as their opportunities would per- mit. There are men, too, of the most recent classes, and men of the years between. A few of them have been in close rela- tions with the Faculty of Applied Science, knowing its accom- plishments, obstacles, and requirements. In all, there is a find of subjects for discussion concerning the old "School" which, in addition to the current engineering topics, would create suffi- cient interest to assure the success of an organization meeting.

We would be glad to hear of such a worthy step being taken.

On another page appears a word from the Secretary of the Faculty in the interests of the employment bureau. That this department is receiving ample recognition from engineers and manufacturers is evidenced by the fact that at no time during the past year were names of applicants for positions on file for any lengthy period. Oftentimes, solicitations for men from manu- facturing firms could not be met. We feel that to the Alumni is due no little credit for this favorable condition of affairs.

The Employ- ment Bureau.

Just at present, however, the bureau is besieged with appli- cants for work—men who, in the press of preparation for exam- inations, have been unable to make their own arrangements for the coming summer. Any information, therefore, that the grad- uate readers of Applied Science may be able to give in the plac- ing of these men during the next fortnight or month, or that will in any way alleviate the burden upon the employment bureau, will be very much appreciated, especially by the undergraduates themselves.

C. Cooper, '99, is engaged at Keokuk, Iowa, with the Keokuk and Hamilton Water Power Company.

J. Hemphill, '09, who for the past two years has been in the employ of the Lake Superior Corporation at Sault Ste. Marie, has accepted a position with the Canton Electrical Company, of Can- ton, Ohio.

PROFESSOR HAULTAIN AT THE DINNER OF THE UNIVERSITY OF TORONTO CLUB OF NEW YORK.

The annual dinner of the University of Toronto Club of New York was held in the Engineers' Club on February 18th. Dr. Reeve, ex-Dean of the Faculty of Medicine, and Professor Haultain were the guests from Toronto.

In his speech Professor Haultain explained that he had not yet been completely absorbed into the academic life; that in many things he still felt very much as an outside graduate, still feeling his way in his new environment. He expressed his appreciation of being a member of the Faculty of Applied Science and Engineering, and touched on matters of University organization from the point of view of the newcomer. After an expression of a very warm appreciation of Dean Galbraith* he spoke in part as follows.

"This morning I had the good fortune to meet one of the prominent members of the Board of Trustees of Columbia University. This Board of Trustees largely combine the functions of our Board of Governors and of our Senate. I told him of my impending speech this evening, and asked him for a word. He said: 'Tell them this; the best thing Columbia ever did was when they put Alumni representatives on the Board of Trustees.'

"Now, as you are aware, representatives of the Alumni, chosen by the Alumni, sit on our Senate. The Senate is a part of the University to which I have not yet penetrated. The other professors, in Arts, and Applied Science and Engineering, have a seat there, but the department of Mining Engineering is not represented. They tell me it is an august body, and is venerated as such. As far as the Engineering Faculty is concerned, its confidence in our Faculty Council seems to leave it without further function than that of confirming our actions. It has the steadying effect of an upper house, but the real work of organization and of progress is done in our own Faculty Council, dependent always, of course, upon the Board of Governors, who hold the purse strings. Here is where the graduates should be, on the Faculty Council and on the Board of Governors. In connection with this there is one thing that has been impressed upon me very strongly in my academic work. I was aware of it when I entered, and it has been impressed upon me at every turn since. The academic work is and must be very different from the field work. There are certain things in which they approach, in all things the one must be governed by the needs of the other, but they are and must remain inherently different. The corollary to this is: The graduate who would be of much service to the University must be in close touch with things academic. The place for him is in the heart of things, the Faculty Council. An irresponsible man, a man who jumped to conclusions without

*See University Monthly, April, 1911

careful study, might do very much harm in such a position, as he would in any other delicate position, but the graduates can be depended upon to choose good men. I would leave an idea with you—an idea that I would fain enlarge upon, did time permit. A graduate on the Faculty Council or on the Board of Governors might have a very valuable catalytic action. There might be more in the results he brought about between others than in what he did himself."

WHAT OUR GRADUATES ARE DOING.

H. F. Shearer, '08, formerly with the Allis Chalmers Company in Cincinnati, is with Smith, Kerry and Chace, in their Toronto office.

H. J. Acres, '03, is with the Hydro-Electric Power Commission, with headquarters in Toronto.

R. E. McArthur, '00, is in the construction department of the Canadian Pacific Railway in Montreal.

W. F. Stubbs, '05, is with the Trussed Concrete Steel Company, Walkerville, Ontario.

P. M. Thompson, '07, is in the employ of the American Bridge Company at Ambridge, Pa.

Chas. Flint, '08, is with the MacDonald Engineering Company, of Chicago.

C. Johnston, '06, is district engineer for the Canadian Northern Railway, with headquarters in Toronto.

T. J. McFarlen, '93, is with the Atikokan Iron Company, Port Arthur, Ontario.

Chester B Hamilton, ir., B A.Sc., '06, is managing a new firm under the name of the Hamilton Gear and Machine Co., which will specialize in cut gears. The factory will be located at the corner of Concord and Van Horne Ave.

D. J. Miller, '10, is connected with the Alberta Central Railway, at Red Deer, Alta.

G. H. Richardson, '88, is in Edmonton, Alta., with the Yellowhead Pass Coal and Coke Company.

S. B. Code, '04, is town engineer of Smith Falls.

W. M. Currie, '04, is managing director, Canada Steel Co. Ltd., Hamilton.

Geo. Hogarth, '09, is assistant engineer, Dept. of Public Works, Parliament Buildings, City.

F. C. Jackson, '01, is in partnership at La Tukue, Quebec, under the name of Jackson and Connolly, contractors.

A. W. Lamont, '09, is sales engineer with the Canadian Westinghouse Co., Ltd., in their Winnipeg office .

E. W. Kay, '07, is in the sales department of the Canadian Westinghouse Company, Winnipeg.

C. S. Dundass, '06, is in Lachine, Quebec, with the Dominion Bridge Company.

Index to Advertisers

some of those peculiar and mysterious laws of gravity elaborated in the classroom, a pail of water was suddenly and with a high initial velocity thrust from a cavity in the ceiling and, while only one man received the pail, several welcomed a temporary relief of mud such as was afforded. This slight detail was closely allied to the action of some man who missed his calling, who so neatly and properly cut the beautiful and mud-colored hair of a dignified and lofty freshman. During the tonsorial treatment a refreshing massage of "eau de la muque" was applied to the back, and a few other cursory operations were performed.

One or two of the daily papers were represented at the elections and courteously devoted a few lines to the method of toughness and graft which pervaded. These mild, youthful representatives of the press would have had more to say had they gone "through the mill." The brave men who ventured "through" to vote were rewarded in the usual manner with a corn-cob pipe, a large (!) red package of dangerous-looking straight-cut, and oranges and apples.

Printed programmes of the evening's sport were circulated. The regular contests ran as follows: The fourth year defeated the third year in the tug-of-war, but was finally overcome by the first year, who hopelessly pulled them "all over the lot." The boxing and wrestling and the paper fight all supplied great amusement. In basketball the Juniors defeated the Seniors, but in broomball the Seniors were the victor over the Juniors. However, there were some sensational events not listed. Not the least of these was a real live, energetic, good-natured scrap lasting for a long period. Two of the three men involved were obliged to borrow outfits of overalls, etc., in order to be legitimately clothed to proceed homeward.

This one night, the roughest of the year, at which each man endeavors to look and be his toughest, is, undoubtedly, a fine institution in the faculty. The health and vigor of youth and manhood here show themselves, always rough, never irresponsible, in one great final fling before settling down to the grind.

The executive chosen for the next year indicates strength, and, by what the individual officers-elect promised as candidates, is to be the finest ever. The executive was elected as follows:

President—W. B. McPherson.

1st Vice-President—B. Watts.

2nd Vice-Presidents—Civils and Architects. J. E. Ritchie; Chemists and Miners, K. MacLachlan; Electricals and Mechanicals—G. J. Mickler.

Treasurer—F. Elliott.

Corresponding Secretary—A. McQueen.

4th Year Rep.—W. J. T. Wright (acc.)

3rd Year Rep.—R. F. B. Wood.

2nd Year Rep.—F. C. Mechin.

Recording Secretary—H. A. Heaton.

Curator—R. G. Matthews.

PRESENTATION TO MR. W. J. SMITHER, '04·

A pleasing sequel to the recent Re-Union Dinner of the Graduating Class of 1904, and the Post-Graduate Class of 1905 of the Faculty of Applied Science, was the presentation by his classmates of a gold watch to Mr. W. J. Smither, '04, who for four years has been seriously ill and during the greater part of that time unable to leave his bed. At present Mr. Smither is in the Orthopedic Hospital, 100 Bloor St. west, Toronto, from which he writes the following letter of thanks to those who were concerned in the gift:

Orthopedic Hospital,
Toronto, Feb. 15, 1911.

The Graduates of '04 and Post Graduates of '05·

Dear Classmates,—I take this means of conveying in part my gratitude for the kind and sympathetic manner in which you remembered me at the Re-Union Dinner, and for the beautiful token of our comradeship of student days, which, I know, is but an evidence of your good wishes for my speedy and complete recovery.

Words are inadequate to express my appreciation of your thoughtfulness, but if your pleasure in giving me this beautiful watch be one-half as great as mine in receiving it, you will all know how I feel when I tender you my heartfelt thanks. It will always be a lasting remembrance of the true friendship and sympathy of my old classmates for me in this long illness.

With sincere wishes for your individual success, I am, most sincerely, yours,

W. J. SMITHER, '04·

"Applied Science" extends its sympathy to Mr. W. J. Graham in his bereavement. Mrs. Graham died on Tuesday morning, March 21st, after a lingering illness. Our regrets are on behalf of every graduate and undergraduate of the "School."

DEATH OF M. T. CULBERT, '02·

On Tuesday, March 14th, Mr. M. T. Culbert, B.A.Sc., succumbed to appendicitis in St. Michael's Hospital, Toronto. Mr. Culbert, as manager of the O'Brien Mining Company, Cobalt, was one of the most prominent among mining men, and will be sadly missed by his associates in Northern Ontario, and by all who knew him, professionally or as a classmate.

G. B. ARMSTRONG, '14·

The death was announced of Mr. G. B. Armstrong, a member of the Class '14, who died at his home in Owen Sound a short time ago.

THE CANADIAN CEMENT AND CONCRETE ASSOCIATION.

Regarding the exhibition of the Canadian Cement and Concrete Association, held recently in Toronto, The Toronto Daily Globe, editorially, has this to say:

"The Cement Show, which has been open for several days, is one of the most attractive industrial exhibitions Toronto has ever had. Its drawing power is due not so much to its appeal to the sense of the beautiful as to the growing importance of cement for a great variety of industrial purposes. These have multiplied in number so rapidly in recent years as to be positively bewildering to all but the experts, and even they find it difficult to keep up with the procession."

The Canadian Cement and Concrete Association organized in Toronto in April, 1908, has been patterned after the Concrete Institute of Great Britain, and the National Association of Cement Users of the United States, each of which has established itself in its own country as an authoritative educational body on matters relating to the uses of cement and concrete. Similarly the Canadian Association aims at advancing the knowledge of concrete and reinforced concrete, and directing attention to the uses to which these materials can be best applied. It will shortly issue a volume of proceedings for the year, a feature of which will be its standard specifications. That the Association is not a money-making organization is evidenced by the fact that it costs practically twice as much to conduct the annual exhibition and convention as is received from the price of admission, and the sale of space.

The papers read at the convention were, for the most part, of a superior kind, and the discussions following them were instructive. Mr. Richard L. Humphrey, director of the Federal Testing Laboratories of the United States, contributed two papers, both of which were listened to with much interest. It is probable that the next meeting will be held in Montreal.

––––––––

THE U. OF T. CLUB IN PITTSBURG.

The University of Toronto Club of Pittsburg has a large membership that is constantly growing. The Club, primarily composed of graduates from the Faculty of Applied Science, includes all other faculties of the University that may be represented in Pittsburg or vicinity. At the annual meeting for the election of officers, at the Hotel Henry, a short time ago, the following officers were elected for the ensuing year:

President—H. M. Scheibe.
Vice-President—D. W. Marrs.
Secretary-Treasurer—J. G. R. Alison, 55 Water street, Pittsburg.

THE D. L. S. AND O. L. S. EXAMS.

The following are the names of the men who were successful at the recent Dominion Land Surveyors' examinations: C. E. Bush, 'o7; G. C. Cowper, '07; F. M. Eagleson, '08; A. E. Glover, '09; J. E. Gray, '09; W. J Johnston, '09; R M. Lee; E. S. Martindale, '09; O. W. Martyn, '09; F. V. Seibert, '09; C. M. Walker, '09·

Those successful in the Ontario Land Surveyors' examinations are as follows: R. M. Anderson, '08; H. W. Tate, '09; J. T. Ransom, '08; C. B. Allison, '08; W. E. Taylor, '09; W. G. McGeorge, '08; J. E. Jackson, 'o9; J. A. Brown, '07; A. E. Jupp, '06; R. Grant, '09; graduates of the Faculty of Applied Science, and N. J. Slater, A. McMeekin, A. Roger, and C. H. Attwood, of other universities.

ERRATA.

In the article entitled, "Incandescent Lamps," by M. B. Hastings, in February issue (Vol. IV., No. 4) the last lines of page 128 should read, "At the present time the 'Mazda' lamp is composed of tungsten filaments, but if carbon or some at present unknown filament be found which would result in a lamp of higher efficiency than the present tungsten filament lamp, it would be the 'Mazda.' "

Again, on page 130, the percentages in the first line should read 0.45. and 0.18 respectively, and the fifth, 35 per cent. instead of 3 per cent.

(Mr. Hastings' paper is a valuable one, and we regret the overlooking of this, at the proper time. Readers will confer a favor by pencilling the necessary change.)

WHAT THE GRADUATES ARE DOING.

J. T. Johnston, '08, is assistant engineer, section five, Trent Valley Canal.

T. H. Alison, '92 is Secretary and Chief Engineer of the Bergen Point Iron Works, Bayonne, N. J.

James A. Beatty, '03, is in business in Peterboro' with Morrow and Beatty, general contractors.

E. R. Birchard, '09' is in charge of the drafting and designing for the Canada Producer and Gas Engine Co., Barrie, Ont.

J. H. Kennedy, '88' is chief engineer for the Great Northern Railway, Vancouver, B. C.

H. E. Brandon, '06, is chief engineer for the Vulcan Iron Works, Winnipeg.

H. S. Carpenter, '97' is Superintendent of Highways, Dep't of Public Works, Regina.

LIST

OF

GRADUATES

AND THEIR

BUSINESS ADDRESSES

APPLIED SCIENCE
APRIL, 1911

GRADUATES.

Graduates are requested to inform the Secretary of changes in their addresses.

1881.
1. J. L. MORRIS, C.E., O.L.S. Pembroke, Ont.
   Morris and Moore Land Surveyors and Architects.

1832.
1. D. Jeffrey Windsor, Missouri.
   Contractor.
1. J. H. Kennedy, C.E., O.L.S., Vancouver, B.C.
   Chief Engineer, Great Northern Ry.
1. J McAREE, B.A.Sc., D.T.S. (Deceased.)

1883.
1. D. Burns, O.L.S., A.M., Can. Soc. C.E., Pittsburgh, Pa.
   Instructor in Mathematics and Plan Drawing, Carnegie Technical Schools.
1. G. H. DUGGAN, M. Can. Soc. C.E., Glace Bay, N.S.
   Dominion Coal Co., Ltd.
1. J. W. TYRRELL, C. E., D.L.S., Hamilton, Ont·
   Tyrrell & MacKay, Consulting Engineers and Surveyors.

1884.
1. W. C. KIRKLAND, New Orleans, La.
   Principal Assistant Engineer, Drainage, Sewage and Water Board of New Orleans.
1. J. McDOUGALL, B.A. (Deceased.)
1. A. R. Raymer, Pittsburgh, Pa.
   Assistant Chief Engineer, P. & L. E. Ry.
1. JAMES ROBERTSON, O.L.S., Toronto, Ont.
   Commissioner, The Canada Co.
1. E. W. STERN, M. Am. Soc. C.E., 103 Park Ave., New York.
   Consulting Civil Engineer.

1885.
1. J. F. BLEAKLEY, Bowmanville, Ont.
   Civil Engineer.
1. H. J. BOWMAN, D. & O.L.S., M. Can. Soc. C.E., Berlin, Ont.
   Bowman & Connor.
1. E. E. HENDERSON, O.L.S., Henderson, P.O., Me
   Civil Engineer.
1. B. A. LUDGATE, O.L.S., Pittsburgh, Pa.
   Assistant Engineer, P. & L. E. Ry.
1. O. McKAY, O.L.S., Walkerville, Ont.
   Civil Engineer and Surveyor.

1886.

1. A. M. BOWMAN, D.L.S.,   Pittsburgh, Pa.
   Pennsylvania Contracting Co.
1. E. B. HERMON, D. & O.L.S.;   Vancouver, B.C.
   Assistant Engineer Vancouver Power Co.
1. ROBERT LAIRD, O.L.S.,   Haileybury, Ont.
   Laird & Routly, Engineers and Surveyors.
1. T. KENNARD THOMSON C.E., M. Can. Soc. C.E., M. Am. Soc. C.E.,
   Hudson Terminal Building, New York.
   Consulting Engineer.
1. H. G. TYRRELL, C.E., A.M. Can. Soc. C.E., 2151 Fulton Ave., Cin., Ohio.
   Chief Engineer, The Brackett Bridge Co.

1887.

1. J. C. Burns (deceased.)
1. A. E. LOTT,   Los Angeles, Cal.
   Consulting Railway Engineer, 441 Bradbury Building.
1. A. L. McCULLOUGH, O.L.S., B.C.L.S., A.M. Can. Soc. C.E., Nelson, B.C.
   Engineer and Surveyor.
1. F. MARTIN, M.B., O.LS.,
   Physician.
1. C. H. PINHEY, D. & O.L.S.,   110 Wellington St., Ottawa, Ont.
1. J. ROGERS, O.L.S.,   Mitchell, Ont.
   Town Engineer.

1888.

1. J. F. APSEY, O.L.S.,   3205 Wallbrook Ave., Baltimore, Md.
   Assistant Divison Engineer, Baltimore Sewerage Commission.
1. W. T. ASHBRIDGE, C.E.,   1444 Queen St. E., Toronto, Ont.
   Engineer and Surveyor.
1. EDWARD F. BALL, A.M. Can. Soc. C.E. ,
   335 Madison Ave., New York, N.Y.
   *Chief Assistant Engineer, Land and Tax Department, N. Y. Central &
   Hudson River Railroad.*
1. D. B. BROWN, O.L.S.,   Quebec, P.Q.
   Locating Engineer, Transcontinental Ry. (G.T.P.)
1. C. M. CANNIFF,   Toronto, Ont.
   Fielding & Canniff Co., Consulting Engineers.
1. H. J. CHEWETT, B.A.Sc., C.E., A.M. Can. Soc. C.E.,
   Manning Arcade, Toronto, Ont.
   Mechanical Engineer, Evans Rotary Engine Co., Ltd.
1. J. GIBBONS, D. & O.L.S.,   Ottawa, Ont.
   Surveying Staff, Department of Interior.
1. R. McDOWALL, O.L.S., C.E.,, A.M. Can. Soc. C.E ,
   Town Engineer.   Owen Sound, Ont.
1. G. W. McFARLEN, O.L.S.,   Toronto, Ont.
   City Engineer's Staff.
1. C. J. MARANI,   Anacortes, Wash.
   Designing and Consulting Structural Engineer for the Russia Cement Co
1. G. R. MICKLE, B.A.,   Toronto, Ont.
   Mine Assessor, Province of Ontario.

1888—*Continued.*

1. J. H. MOORE, O.L.S.,   Smith's Falls, Ont.
Town Engineer.
1. G. H. RICHARDSON,   Edmonton, Alta.
Managing Director Yellowhead Pass Coal & Coke Co.
1. K. ROSE,   Curry Bldg., Toronto.
Manager, Evans Rotary Engine Co. of Canada.
1. J. E. ROSS, D. & O.L.S,   Kamloops, B.C.
Surveying Staff, Department of Interior.
1. C. H. C. WRIGHT, B.A.Sc.,   Toronto, Ont.
Professor of Architecture, University of Toronto.

1889.

1. B. CAREY,   Toronto, Ont.
1. W. J. CHALMERS,   Pittsburgh, Pa.
With Vanport Beaver Co.
1. W. A. CLEMENT, M. Can. Soc. C.E.,   Vancouver, B.C.
City Engineer.
1. G. F. HANNING,   Toronto, Ont.
1. H. E. T. HAULTAIN, C.E , M. Can. Soc. C.E.,   Toronto, Ont.
Professor of Mining Engineering, University of Toronto.
1. J. IRVINE,   Vancouver, B.C.
Locating Engineer, C.N.R.
1. D. D. JAMES, B.A., B.A.Sc.,   227 George St., Toronto.
Surveyor.
1. F. X. MILL, (deceased.)
1. H. K. MOBERLEY, D. & S.L.S.,   Moosomin, Sask.
District Engineer and Surveyor.
1. T. R. ROSEBRUGH, M.A.   Toronto, Ont.
Professor of Electrical Engineering, University of Toronto.
1. T. WICKETT, M.D.,   362 Cannon St. E., Hamilton, Ont.
Physician.

1890.

5. W. E. BOUSTEAD, (deceased).
1. F. M. BOWMAN, O.L.S., C.E.,   Pittsburgh, Pa.
Secretary & Structural Engineer, Riter-Conley Mfg. Co
1. M. A. BUCKE, M.E. (deceased).
1. G. D. CORRIGAN, (deceased).
1. J. A. DUFF, B.A., (deceased).
1. A. B. ENGLISH, (deceased).
1. N. L. GARLAND   76 Wellington St. W., Toronto, Ont.
1. J. HUTCHEON, O.L.S.,   Guelph, Ont.
Engineer and Surveyor.
1. W. L. INNES, O.L.S., C.E.,   Simcoe, Ont.
Manager, Dominion Canners, Ltd.
1. E. B. MERRILL, B.A., B.A.Sc., M. Can. Soc. C.E., M. Am. Inst. E.E.
Toronto, Ont.
Consulting Engineer, Toronto General Trusts Building.

1890—*Continued.*

1. J. R. PEDDER (deceased).
3. R. A. ROSS, E.E.   80 St. Francois Xavier.St., Montreal, Que.
   Ross & Holgate, Consulting Electrical and Mechanical Engineers.
1. T. H. WIGGINS, O.L.S.,   Saskatoon, Sask.
   Civil Engineer and Dom. Land Surveyor.
1. W. J. WITHROW,   Ottawa, Ont.
   Patent Examiner, Patent Office.

1891.

1. H. J. BEATTY, O.L.S.,   Eganville, Ont.
   Engineer and Surveyor.
1. T. R. DEACON, O.L.S., M. Can. Soc. C.E.,   Winnipeg, Man.
   President and General Manager, Manitoba Bridge & Iron Works,Ltd.
1. C. W. DILL, M. Can. Soc. C.E.,   Toronto, Ont.
   *C. W. Dill & Co., Civil Engineers and Contractors, 318 Continental
   Life Building.*
5. O. S. JAMES, B.A.Sc.,   227 George St., Toronto, Ont·
1. A. LANE (deceased).
1. J. E. McALLISTER, B.A.Sc., C.E.,   Greenwood. B,C.
   Manager, British Columbia Copper Co., Ltd.
3. E. B. MERRILL, B.A., B.A.Sc., M. Can. Soc. C.E., M. Am. Inst. E.E.
                                              Toronto, Ont.
   Consulting Engineer, Toronto General Trusts Building.
1. J. E. A. MOORE, C.E.,   10074 Kee Mar Court, Cleveland, O.
   Consulting and Contracting Engineer.
1. W. NEWMAN, O.L.S., A.M. Can. Soc. C.E.,   Windsor, Ont·
   Consulting Engineer.
1. J.K. ROBINSON (deceased).
1. W. B. RUSSEL,   318 Continental Life Bldg., Toronto, Ont.
   Civil Engineer and Contractor.
1. G. E. SILVESTER, O.L.S., M. Am. Inst. M.E.,   Copper Cliff, Ont.
   Chief Engineer, Canadian Copper Co.
1. H. D. SYMMES,   Niagara Falls S., Ont.
   Engineer and Contractor.

1892.

1. J. R. ALLAN, O.L.S.,   Macleod, Alta.
   Ranchman.
1. T. H. ALISON, B.A.Sc., C.E.,   Bayonne, N.J.
   Secretary and Chief Engineer, Bergen Point Iron Works.
1. A. G. ANDERSON,   Port Dover, Ont.
   Hardware Merchant.
1. C. FAIRCHILD, D. & O.L.S.,   Brantford, Ont.
   Surveying Staff, Department of Interior.
1. J. B. GOODWIN, B.A.Sc.,   Gasport, N.Y.
   Superintendent of Construction, Empire Eng. Corporation of New York.
4. C. E. LANGLEY,   Continental Life Bldg., Toronto, Ont.
   Langley & Howland, Architects.

1892—*Continued.*

1. A. T. LAING, B.A.Sc., Toronto, Ont.
Secretary, Faculty of Applied Science, University of Toronto.

1. E. J. LASCHINGER, B.A.Sc., M.E., Johannesburg, Transvaal S.A.
Mechanical Engineer, H. Eckstein & Co.

5. W. L. LAWSON, B.A.Sc, Stirling, Col.
Manager, Stirling Brush & Fort Morgan Co.

3. W. A. LEE, B.A.Sc. (deceased).

1. B. McENTEE, B.A.Sc., 28 Queen St. E., Toronto.
Stationer.

3. C. G. MILNE, B.A.Sc. (deceased).

1. CHAS. H. MITCHELL, B.A.Sc., C.E., M. Can. Soc. C.E., M. Am. Soc. C.E.
Consulting Hydro-Electric Engineer, Trader's Bank Bldg.. Toronto.

1. N. L. PLAYFAIR, Midland, Ont.
Superintendent Playfair Lumber Co.

1. J. M. PRENTICE (deceased).

1. J. A. ROSS, Cleveland, Ohio.
Designer L. S. & M. S. Railway, Engineering Office.

1. ALBERT N. SMITH, Youngstown, Ohio.
Engineer, Wm. B. Pollock Co.

1. R. W. THOMSON, B.A.Sc., M.E., 4 Charles St., Toronto, Ont.
Mining Engineer.

3. A. V. WHITE, M.E., Toronto, Ont.
Mechanical Engineer.

1893.

1. A. G. ARDAGH, Barrie, Ont.
Land Surveyor and Civil Engineer.

.*H. F. BALLANTYNE, B.A.Sc. 244 Fifth Ave., New York, N.Y.
Architect.

1. G. L. BROWN, O.L.S., A.M. Can. Soc. C.E., Morrisburg. Ont.
Civil Engineer and Land Surveyor.

1.*L. C. CHARLESWORTH, D.L.S.. Edmonton, Alta.
Director of Surveys for Alberta.

1. T. H. DUNN, O.L.S. Winchester, Ont.
Engineer and Surveyor.

1. J. M. R. FAIRBAIRN, P.L.S., Westmount, Que.
Assistant Engineer, C.P.R.

4.*W. FINGLAND, 334 Portage Ave.. Winnipeg, Man.
Architect and Structural Engineer.

1. C. FORRESTER Toronto, Ont.

1.*WALTER J. FRANCIS, C E. M. Can. Soc. C.E,. M. Am. Soc. C.E.
Montreal, Que.
Consulting Engineer, 28 Commercial Union Building.

3.*A. R. GOLDIE, Galt, Ont.
Manager, Goldie & McCulloch Co.

3. S. C. HANLY, Midland. Ont.
Midland Engine Works Co.

4.*J. KEELE. B.A.Sc., Ottawa, Ont.
Geological Survey of Canada.

---

*Diploma with honours.

1893—*Continued.*

1. J. T. Laidlaw, B.A.Sc., M.E.,   Cranbrook, B.C.
   Consulting Mining Engineer.
3. F. L. Lash,   Bandoeng, Java.
   Manager, Electrical Supply Co., Board of Trade Building.
1. A. L. McAllister, B.A.Sc., 612 Continental Life Bldg., Toronto, Ont.
   Consulting Engineer.
1. T. J. McFarlen,   80 Waverley Rd., Toronto, Ont.
   Chemist.
1.*A. J. McPherson, B.A.Sc., D.L.S.,   Regina, Sask.
   Superintendent of Highways, Province of Saskatchewan.
1. A. F. Macallum, B.A.Sc., C.E..   Hamilton, Ont.
   City Engineer.
1. W. T. Main,   Wells St. Depot, Chicago, Ill.
   Division Engineer, C. &. N. W. Ry.
1. V. G. Marani,   Cleveland, Ohio.
   City Building Inspector, City Hall.
1. W. Mines. B.A.Sc.,   Cleveland, Ohio.
   With Brown Hoisting Co.
3.*J. M. Robertson,   Montreal, P.Q.
   *Superintendent Repair and Testing Department, Montreal Light. Heat
   and Power Co.*
1. R. Russell,   Pembroke, Ont.
   Railway Contractor.
1.*F. N. Speller, B.A. Sc.,   Pittsburgh, Pa.
   Metallurgical Engineer, National Tube Co.
1. R. H. Squire, B.A. Sc., O.L.S (deceased).
1. W. V. Taylor, O.L.S., A.M.Can. Soc. C.E.,   Quebec, P.Q.
   Quebec Harbour Commissioners.
1.*R. B. Watson,   Regina, Sask.
   Department of Public Works.

1894.

3.*R. W. Angus, B.A.Sc.,   Toronto, Ont.
   Professor of Mechanical Engineering, University of Toronto.
1. H. F. Barker,   Box 31, Halifax, N.S.
1. A. T. Beauregard, B.A.Sc.,   East Orange, N.J.
   Laboratory Engineer, Public Service Corporation of New Jersey.
1. A. E. Bergey,   Pittsburgh, Pa.
   Carnegie Technical School.
3. D. G. Boyd,   Toronto, Ont.
   Draftsman, Public Works Department.
3. W. A. Bucke,   Toronto, Ont.
   District Manager, Canadian General Electric Co.
1. J. Chalmers, O.L.S., A. M. Can. Soc. C.E.,   Edmonton, Alta.
   Structural Engineer, Department of Public Works.
4.*J. A. Ewart. B.A.Sc.,   193 Sparks St., Ottawa, Ont.
   Architect and Engineer.

*Diploma with honours.

1894—*Continued.*

3. W. J. HERALD, B.A.Sc.,   Toronto, Ont.
   Engineering Department, Canada Foundry Co.
;. H. E. JOB, B.A.Sc.,   Hamilton, Ont.
   Manufacturer of Electrical Machinery and Apparatus.
3. A. C. JOHNSTON, B.A.Sc., M.E.,   Philadelphia, Pa.
   Vice-President and Chief Engineer, The J. M. Dodge Company.
1. S. M. JOHNSTON, B.A.Sc., P.L.S.,   Greenwood, B C.
   City Engineer.
1. J. E. JONES,   Engineers' Club, New York, N.Y.
3. N. M. LASH,   Montreal, P.Q.
   Assistant Electrical Engineer, Bell Telephone Co.
1.*A. L. McTAGGART, B.A.Sc.,   Rockefeller Bldg., Cleveland O
   Office of A. G. McKee, Consulting Engineer.
3.*W. MINTY, B.A.Sc.,   Blackburn. Eng.
   With Messrs. Yates & Thom. Ltd , Engineers.
3. C. J. NICHOLSON,   Hamilton, Ont.
   ' *Assistant Engineer, Hamilton, Guelph & Waterlo ' Ry.*
1. H. ROLPH,   Montreal, Que.
   Secretary, John S. Metcalf Co., Ltd.
1. J. D. SHIELDS, B.A.Sc.,   Toronto, Ont.
   Sewer Engineer, Staff of City Engineer.
1. ANGUS SMITH, O.LS., A.M. Can. Soc. C.E.,   Regina, Sask.
   City Engineer.
3. A. K. SPOTTON,   Galt, Ont.
   Chief Engineer, Goldie & McCulloch Engine Works.
3. R. T. WRIGHT, B.A.Sc.,   East Pittsburgh, Pa
   Engineering Department, Westinghouse Machine Co.

1895.

1. J. ARMSTRONG, B.A.Sc.,   Quebec, Que.
   District Engineer, G.T.P. Ry.
3. A. E. BLACKWOOD,   30 Church St., New York.
   Manager New York Office, Sullivan Machinery Co.
1. E. J. BOSWELL, D.L.S.,   Winnipeg, Man.
3. G. BREBNER (deceased).
3. W. M. BRODIE, B.A.Sc.,   Pittsburgh, Pa.
   With the Green Engineering Co. of Chicago.
3. L. I. BROWN,   115 Broadway, New York.
   Superintendent, The Foundation Co.
4. R. J. CAMPBELL,   Chicago. Ill.
   Artist, Chicago Tribune.
3. A. W. CONNOR, B.A., C.E.,   36 Toronto St., Toronto, Ont.
   Bowman & Connor, Consulting Engineers.
1. J. S. DOBIE, B.A.Sc., O. & D.L.S.,   Thessalon, Ont.
1. F. W. GUERNSEY,   Bankhead, Alta
   Assistant General Manager, Bankhead Mines, Ltd.

---

*Diploma with honours.

1895.—Continued.

4.*A. H. HARKNESS, B.A.Sc.,   Toronto, Ont.
 Structural Engineer, Confederation Life Building.

3. H. S. HULL, B.A.Sc.,   Johnstown. Pa.
 Structural Drawing, Cambria Steel Co.

3.*J. McGOWAN, B.A., B.A.Sc.,   Toronto, Ont.
 Associate Professor of Applied Mecianics, University of Toronto.

3. W. N. McKAY,   Georgetown, Ont .
 Manager, Bank of Hamilton.

3. H. L. McKINNON, B.A.Sc.,   Cleveland, Ohio.
 Vice-President of The C. O. Bartlett & Snow Co.

1. W. W. MEADOWS, D. & O.L.S.,   Maple Creek, Sask.
 Department of Public Works.

1. F. J. ROBINSON, D. & O.L.S.,   Regina, Sask·
 Deputy Minister of Public Works. Saskatchewan.

3. F. T. STOCKING,   Toronto, Ont.
 Hydro-Electric Commission. '

3. R. C. C. TREMAINE, B.A.Sc. (Deceased).

1896.

2.*J. W. BAIN, B.A.Sc.,   Toronto, Ont.
 Associate Professor of Applied Chemistry, University of Toronto.

2. L. T. BURWASH,   Dawson, Y.T.
 Mining Recorder.

3.*G. M. CAMPBELL,   Riverside, Ill.
 Superintendent, Power Apparatus Shops, Western Electric Co.

2. J. A. DECEW, B.A.Sc.,   615 Canadian Express Bldg., Montreal, Que.
 Consulting Chemical Engineer.

3.*H. P. ELLIOTT, B.A.Sc., E.E.,   Pittsburgʰ, Pa.
 Electrical Engineer, Westinghouse Electric and Manufacturing Co.

3. W. C. GURNEY,   Toronto, Ont.
 Vice-President, Gurney Foundry Co., Ltl.

3.*H. V. HAIGHT, B.A.Sc.,   Sherbrooke, P.Q.
 Chief Engineer, Canadian Rand Drill Co.

1. W. F. LAING, (deceased).

3. R.R. LAWRIE, (deceased).

3. C. MacBETH, B.A Sc., (deceased).]

3. J. A. MacMURCHY,   Pittsburgh, Pa.
 Chief Draftsman, Turbine Department, Westinghouse Machine Co.

1. T. MARTIN, B.A.Sc.,   Calgary, Alta.
 Assistant Divisional Engineer. C.P.R., Western Division.

3. R.R. SCHEIBE,   Toronto, Ont.
 Sales Manager, Brigdens Ltd.

1897.

2. E. ANDREWS, B.Sc., A.M.I., C.E.,   Portmadoc, N. Wales.
 Resident Engineer, Maenofferen Slate Quarry Co., Limited.

2.*J. A. BOW,   Great Falls, Mont.
 B. & M. Smelter.

———

*Diploma with honours.

1897—*Continued.*

1. H. S. CARPENTER, B.A.Sc., O.L.S.,   Regina, Sask.
Superintendent of Highways, Department of Public Works.
5. H. W. CHARLTON, B.A.Sc.,   Ottawa, Ont.
Assistant Chemist at Experimental Farm.
4.*E. A. FORWARD, A.M. Can. Soc. C.E.,   Lockport, Man.
Engineer-in-charge, St. Andrew's Lock and Dam.
3.*A. T. GRAY, B.A.Sc.,   Schenectady, N.Y.
Designing Engineer on Steam Turbines, General Electric Co.
3. W. A. B. HICKS,   Buffalo, N.Y.
With Lackawanna Steel Co.
4. C. F. KING,   Toronto, Ont.
Rep. of Mortimer Co. of Ottawa.
1. H. W. PROUDFOOT, (deceased).
2.*A. H. A. ROBINSON, B.A.Sc., M.A.I., M.E.,   Haileybury, Ont.
Mine Inspector.
4. W. F. SCOTT,   Toronto, Ont.
Structural Engineer and Consulting Architect.
3.*W. R. SMILEY, B.A.Sc.,   Cleveland, Ohio.
With Wellman-Seaver-Morgan Engineering Co
2.*W.W. STULL, B.A.Sc., O.L.S.,   Sudbury, Ont.
Surveyor and Mining Engineer.
1.*M. B. WEEKES, B.A.Sc., D.L.S.,   Regina, Sask.
Department of Public Works.
1. E. A. WELDON,   Winnipeg, Man.
Provincial Land Surveyor's Office.

1898.

1. W. H. BOYD, B.A.Sc.,   Ottawa, Ont.
Geological Survey of Canada.
2. W. E. H. CARTER, B.A.Sc.,   Toronto, Ont.
Consulting Mining Engineer, 83 &85 Front Street, East.
3. E. H. DARLING, A.M. Can. Soc. C.E.,   Hamilton, Ont.
Assistant Engineer, Hamilton Bridge Works Co.
1. W. F. GRANT, B.A.Sc.,   Sault Ste. Marie, Ont.
City Engineer.
1. J. S. KORMANN, B.A.Sc.,   Toronto, Ont.
Manager, Kormann Brewing, Ltd.
3. J. E. LAVROCK,   Vancouver, B.C.
Draftsman, Hermon & Burwell.
4. D. MacKINTOSH, B.A.Sc., B. Arch.,   New York, N.Y.
Chief Superintendent F. M. Andrews, & Co , Metropolitan Tower.
1. F W. McNAUGHTON, O.L.S.,   Winnipeg, Man.
Deputy Minister of Public Works.
1. J. H. SHAW, O.L.S.,   North Bay, Ont.
Surveyor and Engineer.
3. A. E. SHIPLEY, B.A.Sc.,   Nelson, B.C.
Manager, Nelson Coke & Gas Co.

*Diploma with honours.

1898—*Continued*.

3.*F. C. SMALLPIECE, B.A.Sc.,   Montreal, Que.
   Assistant Manager, Canadian General Electric Co.
1. R. W. SMITH, P.L.S.,   Revelstoke, B.C.
   Surveyor.
1.*J. A. STEWART, M.A.,   Hamilton, Ont.
   Engineer and Contractor, 67 Federal Life Building.
1.*H. L. VERCOE,   Montreal, Que.
   Chief Draftsman, Grand Trunk Pacific Ry.
3. T. A. WILKINSON,   New York, N.Y.
   Statistician, Westinghouse Church Kerr Co.
3. D.A. WILLIAMSON, B.A.Sc.,   Hamilton, Ont.
   With Hamilton Bridge Works Co.

1899.

3.*T. BARBER,   Meaford, Ont.
   Hydraulic Engineer, Chas. Barber & Sons.
2. J. T. M. BURNSIDE, B.A.Sc.   Montreal, Que.
3. L. B. CHUBBUCK, B.A.Sc., E.E.,   Hamilton, Ont.
   Engineer, Canadian Westinghouse Co.
2. G. A. CLOTHIER,   Rossland, B.C.
   Engineer, Le Roy Mining Co.
1. C. COOPER,   Keokuk, Iowa.
   Koekuk & Hamilton Water Power Co.
2. R. W. COULTHARD, B·A.Sc.,   Blairmore, Alta.
   General Manager, West Canadian Collieries, Ltd.
3. J. A. CRAIG, B.A.Sc.,   Toronto, Ont.
   Office of Willis Chipman, C.E.
2. J. C. ELLIOTT,   Kelso, Ont.
3. W. E. FOREMAN, B.A.Sc.,   Pittsburgh, Pa.
   Construction Dept., Westinghouse Electric and Mfg. Co.
3. E. GUY, B.A.Sc.,   Industry, Pa.
   Engineering Dept., Westinghouse Electric and Mfg. Co.
3.*W. ALMON HARE, B.A.Sc., A.M. Can. Soc. C.E., Toronto, Ont.
   Secy-Treas. and Chief Engineer, The Standard Engineering Co.
1. R. LATHAM, B.A.Sc.,   Hamilton, Ont.
   Chief Engineer, T. H. & B. Ry.
3. W. MONDS, B.A.Sc.,   36 Toronto St., Toronto, Ont.
   Clark & Monds, Consulting Engineers.
1. J. PATTERSON, B.A.,   Toronto, Ont.
   Physicist, Dominion Observatory.
3. A. S. H. POPE, B.A.Sc.,   Portland Oregon.
   Pope & Wilcox, Electrical & Mechanical Engineers.
2.*G. E. REVELL, B.A.Sc.,   Nelson, B.C.
3.*E. RICHARDS, B.A.Sc.,   Toronto, Ont.
   Assistant Electrical Engineer, City of Toronto.
3. G. A. SAUNDERS,   Wilkinsburg, Pa.
   With Westinghouse Electric & Manufacturing Co.

*Diploma with honours.

1899—*Continued.*

1.*T. SHANKS, B.A.Sc., D.L.S.,   Ottawa, Ont.
Topographical Surveys Branch, Department of the Interior.

1.*D. C. TENNANT, B.A.Sc.,   Lachine Locks, Que.
With Dominion Bridge Co.

3. W. W. VanEVERY,   Sault Ste Marie, Ont.
Soo Corporation.

3. W. E. WAGNER, B.A.Sc.,   Toronto, Ont.
With Toronto Ferry Co.

2. G. H. WATT, D.L.S.,   Ottawa, Ont.
Dominion Land Surveyor.

3. E. YEATES,   Hamilton, Ont.
Manager, London Machine Tool Co., Ltd.

1900.

1. J. L. ALLAN, A.M. Can. Soc., C.E.,   Halifax, N.S·
Office of Provincial Engineer.

2. E. G. R. ARDAGH, B.A.Sc.,   Toronto, Ont.
Lecturer in Applied Chemistry, University of Toronto.

3. J. A. BAIN,   Ottawa, Ont.
Structural Engineer, Dept. of Public Works of Canada.

3. J. H. BARLEY, B.A.Sc.,   Hamilton, Ont.
Canadian Westinghouse Co.

2.*M. C. BOSWELL, M.A., Ph.D.,   Toronto, Ont.
Lecturer in Organic Chemistry, University of Toronto.

1. L. T. BRAY, D. & O.L.S.,   Edmonton, Alta.
District Engineer and Surveyor.

3. J. CLARK,   Pittsburgh, Pa.
Electrician, P. & L. E. R. R.

2. J. E. DAVISON, B.A.Sc.,   Fort William.
Engineering Staff, Can. Northern Ry.

3. E. D. DICKINSON,   Schenectady, N.Y.
With General Electric Co.

3. G. W. DICKSON, B.A.Sc.,   Toronto, Ont.
With Smith, Kerry & Chace.

2.*H.A.DIXON, B.A.Sc., M.L.S.,   Winnipeg, Man.
Division Engineer, Canadian Northern Railway.

2. C. H. FULLERTON, O.L.S.,   New Liskeard, Ont.
Engineer and Surveyor.

3. W. S. GUEST, B.A.Sc.,   Toronto, Ont.
Demonstrator in Electrical Engineering, University of Toronto.

3. W. HEMPHILL B.A.Sc., E.E.,   Buffalo, N. Y.
Superintendent, Cataract Power & Conduit Co.

3. S. E. M. HENDERSON,   422 Rubidge St., Peterboro, Ont.
Designing Engineer, Canadian General Electric Co.

3. J. A. HENRY,   Schenectady, N. Y.
Designing Engineer, General Electric Co.,

2. H. S. HOLCROFT, B. A.Sc., D.L.S.
Surveyor, Peace River District.

---

*Diploma with honours.

1900—_Continued._

3. H. A. JOHNSON.   148 Clinton St., Toronto, Ont.
    Manager, Johnston Oil Engine Co., Limited.

3. J. C. JOHNSTON,   Boston, Mass.
    Plant Inspector, Warren Bituminous Paving Co.

2.*J. A. JOHNSTON, B.A.Sc.,   Ignace, Ont.
    Contractor.

2. R. E. MCARTHUR,   Montreal, Que.
    Const. Dept., C.P.R.

2. J. G. MCMILLAN, B.A.Sc.,   39 Wood St., Toronto, Ont.
    Mining Engineer.

3. L. HAUN MILLER,   Cleveland, Ohio.
    Sales Agent, Bethlehem Steel Co.

2. E. V. NEELANDS, B.A.Sc.,   Cobalt, Ont.
    Hargrave Mines.

1.*E. H. PHILLIPS, D.L.S.,   Saskatoon, Sask.
    Phillips & Phillips, Civil Engineers and Surveyors.

2. J. R. ROAF, B.A.Sc.,   Michel, B.C.
    Draftsman, Crow's Nest Pass Coal Co.

3.*C. H. E. ROUNTHWAITE,   Winnipeg, Man.
    Draftsman, G.T.P. Ry.

2. H. W. SAUNDERS, B.A.Sc.,   Gary, W. Va.
    Division Engineer, U.S. Coal & Coke Co.

1. A. TAYLOR, D.L.S. & M.L.S..   Portage La Prairie, Man.
    Engineer and Surveyor.

1. W. C. TENNANT, B.A.Sc. (deceased).

2. S. M. THORNE, B.A.Sc.,   Cobalt, Ont.
    Engineering Staff, Silver Leaf Mine.

1. F. W. THOROLD, B.A.Sc.,   Toronto, Ont.
    Assistant City Engineer on Const. of Sewage System.

1. H. M. WEIR, B.A.Sc.,   Pachuca, Mex.
    With Real Del Morte Co.

3. F. D. WITHROW,   Ottawa, Ont.
    Patent Examiner, Dept. of Agriculture.

1901.

1. R. H. BARRETT, B.A.Sc., O.L.S. (deceased).

3. W. G. BEATTY,   Fergus, Ont.
    Manager, Beatty Bros., Implement Manufacturers.

3. G. M. BERTRAM,   Joplin, Mo.
    Representative of the Sullivan Machinery Co.

3. W. J. BOWERS (deceased).

3. E. T. J. BRANDON, B.A.Sc.,   Toronto, Ont.
    Assistant Engineer Hydro-Electric Power Comm.

3. W. P. BRERETON, B.A.Sc.,   Winnipeg, Man.
    Assistant Engineer, Power Construction Dept.

*Diploma with honours.

1901—*Continued.*

3. J. T. BROUGHTON, Scottdale, Pa.
Chief Engineer, Scottdale Foundry & Machine Co.

3.*W. G. CHACE, B.A.Sc., Carnegie Library, Winnipeg Man.
Firm of Smith, Kerry & Chace.

3. A. G. CHRISTIE, 1713 Munro St., Madison, Wis.
Professor of Steam Engineering, University of Wisconsin.

3. J. R. COCKBURN, B.A.Sc., A. M. Can. Soc. C.E. Toronto, Ont.
Lecturer in Descriptive Geometry, University of Toronto.

1. W. A. DUFF, Ottawa, Ont.
Assistant Bridge Engineer, Transcontinental Ry.

2.*D. E. EASON, B.A.Sc., Peterboro', Ont.
Division Engineer, Trent Valley Canal.

1.*S. GAGNE, B.A.Sc (deceased).

3. N. R. GIBSON, B.A.Sc., Winnipeg, Man.
Assistant Engineer, Power Const. Dept.

2. A. T. E. HAMER, Wahnapitae, Ont.
Engineering Staff, Canadian Northern Ry. Co.

1. C. HARVEY, B.A.Sc., D.L.S., Kelowna, B.C.
Consulting Engineer and Surveyor.

2. F. C. JACKSON, La Tuque, Que.
Jackson & Connelly, Contractors, N.T.C. Ry.

3.*A. LAIDLAW, Kansas, City, Mo.
District Manager, Trussed Concrete Steel Co.

3. W. C. LUMBERS, Calgary, Alta.
Engineering Staff, C.P.R.

3. A. C. MACDOUGALL, Massena, N.Y.
Asst. Superintendent, Aluminium Co. of America.

3. A. T. C. MCMASTER, B.A.Sc., Copper Cliff,
Engineer on Design & Construction, The Canadian Copper Co.

1. G. MACMILLAN, Ottawa, Ont.
Topographical Surveys Branch, Dept. of Interior.

3.*H. G. MCVEAN, B.A.Sc., Moose Jaw, Sask.
Contractor and Engineer.

2. W. C. MATHESON, Joliette, Que.
With McKenzie Mann. Co.

3. H. T. MIDDLETON, Englewood Cliffs, N.J.

2. J. L. R. PARSONS,.B.A., D.L.S., Winnipeg, Man.
Engineer and Surveyor.

1. G. H. POWER, North Battleford, Sask.
Western Canada Rep. of Willis Chipman, C.E.

3.*H. W. PRICE, B.A.Sc., Toronto, Cnt.
Lecturer in Electrical Engineering, University of Toronto.

1. H. P .RUST, B.A.Sc., A.M.Can. Soc., C.E., New York, N.Y.
With Messrs. Viele, Blackwell & Buck.

3. M. V. SAUER, B.A.Sc., Niagara Falls, Ont.
Assistant Engineer, Ontario Power Co.

3. W. H. STEVENSON, B.A.Sc., Monadnock Block. Chicago, Ill.
Secretary Power Plant Specialty Co.

*Diploma with honours.

1901—*Continued.*

1. R. D. WILLSON,   Winnipeg, Man.
    Assistant City Engineer.

1902.

3.*H. G. BARBER,   Ottawa, Ont.
    Topographical Surveys Branch, Department of the Interior.
1. W. J. BLAIR, B.A.Sc., D. & O.L.S.,   New Liskeard. Ont.
    Civil Engineer and Surveyor.
3. J. M. BROWN,   Pittsburgh, Pa.
    With Westinghouse Machine Co., Steam Turbine Dept.
2. W. G. CAMPBELL.   Toronto, Ont.
2. A. R. CAMPBELL,   Toronto, Ont.
    Universal Mfg. Co., Ltd., St. James Chambers.
3. C. G. Carmichael (deceased).
2.*W. CHRISTIE, B.A.Sc.,   Prince Albert. Sask.
    Dominion Land Surveyor.
2. F. T. CONLON,   Thorold, Ont.
    Welland Canal Engineering Staff.
3. H. V. CONNOR,   Pittsburgh. Pa.
    With Westinghouse Electric and Mfg. Co.
2.*M. T. CULBERT, deceased.
2. R.CUMMING,   Toronto, Ont.
    General Contractor. 50 Front St. E.
1. W. E. DOUGLAS, B.A.,   Toronto, Ont.
    Contractor, 152 Bay St.
3.*R. J. DUNLOP,   Toronto, Ont.
    With Canadian Westinghouse Co.
2. W. M. EDWARDS, B.A.Sc.,   1510 5th St. West, Calgary, Alta.
    With Smith, Kerry & Chace.
3. W. ELWELL (deceased).
2. J. M. EMPEY, B.A.Sc.,O.L.S.,D.L.S.,   Calgary, Alta,
    District Engineer and Surveyor. Dept. of Public Works.
2.*D. L. H. FORBES,   Yzabal, Sonora, Mex.
    Metallurgical Engineer El Tigre Mining Co.
1.*A. E. GIBSON, B.A.Sc.,   Toronto. Ont.
    Office of Haney & Miller, Engineers and Contractors.
3. A. C. GOODWIN,   Pittsburgh, Pa.
    Draftsman, Aluminium Co. of America.
3. C. P. HENWOOD,   McKeesport, Pa.
    Draftsman, National Tube Co.
3. D. M. JOHNSTON,   London, Ont ·
    Inspector, London Sub-station, Hydro-Electric Power Comm.
2. R. H. KNIGHT, B.A.Sc.,D.L.S.,   Edmonton, Alta.
    Driscoll & Knight, Engineers and Surveyors.
5.*F. L. LANGMUIR, B.A.Sc.,Ph.D.,   Toronto, Ont.
    Chemist, M. Langmuir Mfg. Co.
3. A. H. McBRIDE, B.A.Sc.,   Toronto, Ont.
    Assistant Engineer, Hydro-Electric Power Commission.

*Diploma with honours.

1902—*Continued.*

1. A. L. McLennan, D.L.S.,   Toronto, Ont.
   Office of York Co. Engineer.
3. J. T. Mackay,   Toronto, Ont.
   Student in Faculty of Medicine, University of Toronto.
3. J. F. S. Madden,   Winnipeg, Man.
   Erecting Engineering Dept., Can. Gen. Electric Co.
3.*C. H. Marrs,   Pittsburgh, Pa.
   Designing Engineer, Riter-Conley Mfg. Co.
3. P. Mathison, B.A.Sc.,   Hamilton, Ont.
   Electrical Eng.. Canadian Westinghouse Co.
3. R. S. Mennie,   Pittsburgh, Pa.
   With Crucible Steel Co. of America.
2. H. H. Moore, D.L.S.,A.M. Can. Soc. C.E.,   Calgary, Alta.
   Dominion Land Surveyor and Engineer.
1.*T. S. Nash,   Ottawa, Ont.
   Topographical Surveys Branch, Department of the Interior.
1. G. G. Powell, B.A.Sc.,   Toronto, Ont.
   Asst. City Engineer, Roadways Dept.
1.*W. F. Ratz, D.L.S. (deceased).
3. H. D. Robertson, B.A.Sc.,   Toronto, Ont.
   Miller, Cumming & Robertson, Engineers and Contractors.
3.*D. Sinclair, B.A.Sc. (deceased).
2.*I. J. Steele, D.L.S.,   Ottawa, Ont.
   Topographical Surveys Branch, Dept. of Interior.
3. W. H. Sutherland, B.A.Sc..   107 St. James St.. Montreal, Que.
   Assistant Engineer, Montreal Water and Power Co.
3.*T. F. Taylor,   494 Concord Ave., Toronto, Ont.
2.*C. M. Teasdale,   Concord, Ont.
   Surveyor.
3. A. A. Wanless,   Sydney Mines, N.S.
   Lecturer, Sydney Mines Technical Schools.
3. H. J. Zahn, B.A.Sc.,   235 Calumet St., Detroit, Mich.

1903.

3. H. G. Acres,   Toronto, Ont.
   Asst. Engineer, Hydro-Electric Power Commission.
1. J. G. R. Alison,   Pittsburgh, Pa.
   With Riter-Conley Mfg. Co.
3.*H. H. Angus, B.A.Sc..   Bethlehem, Pa.
   Draftsman, Bethlehem Steel Co.
3. J. A. Beatty,   Peterboro', Ont.
   Morrow & Beatty, Contractors.
3.*J. Breslove,   East Pittsburgh, Pa.
   Steam Turbine Engineer, Westinghouse Machine Co.
2. J. H. Burd, O.L.S.,   Sudbury, Ont.
   Engineer and Surveyor.

*Diploma with honours.

1903—*Continued.*

1.*E. L. BURGESS, D.L.S.,   Ottawa, Ont.
Topographical Surveys Branch, Department of the Interior.

2. N. A. BURWASH, B.A.Sc.,   Whitehorse, Y.T.
Surveyor.

1. F. F. CLARKE, D. & O.L.S. A.M. Can. Soc. C.E.,   Toronto, Ont.
Divisional Engineer, Can. Northern Ry.

2. C. L. COULSON,   Welland, Ont.

3.*A. E. DAVISON, B.A.Sc.,   Toronto, Ont.
Engineering Staff, Hydro-Electric Power Commisssion.

3. C. J. FENSOM. B.A.Sc., M.E.,   Toronto, Ont.
Consulting Mechanical Engineer, 43 Victoria St.

2.*E. O. FUCE, O.L.S.,   Galt, Ont.
Consulting Civil Engineer.

3*.F. A. GABY, B.A.Sc.,   Toronto, Ont.
Assistant Chief Engineer, Hydro-Electric Power Commission.

1. J. C. GARDNER, B.A.Sc.,   Niagara Falls, Ont.
City Engineer.

3. R. E. GEORGE,   Dover, N.H.
Electrical and Gas Engineer, The United Gas & Electric Co.

1.*P. GILLESPIE. B.A.Sc.,   Toronto, Ont.
Lecturer in Theory of Construction, University of Toronto.

1. W. A. GOURLAY,   Toronto. Ont.
Engineering Staff, C.P.R.

2. J. F. HAMILTON, B.A.Sc., C.E.,   Lethbridge, Alta.
Hamilton & Young, Dominion Land Surveyors and Engineers.

2. G. S. HANES. B.A.Sc, O.L.S.   North Vancouver, B.C.
City Engineer.

2. F. Y HARCOURT. B.A..   Port Arthur, Ont.
Engineer, Public Works Dept.

1. L. J. HAYES,   Chicago, Ill.
Structural Engineer, Corn Products Refining Co.

1.*F. D. HENDERSON, Sec'y. Board of Examiners for D.L.S.. Ottawa, Ont.
Topographical Surveys Branch, Department of the Interior.

5.*J. A. HORTON,   New Ontario.

3. J. G. JACKSON,   Toronto Ont.
Electrical Department, City Hall.

3. C. K. JOHNSTON,   Pefferlaw. Ont.
Merchant.

1. H. JOHNSTON, O.L.S.,   Berlin, Ont.
Davis & Johnston, Civil Engineers and Surveyors.

3. A. G. LANG,   Toronto, Ont.
Underground Superintendent, Toronto Hydro-Electric System.

1.*A. J. LATORNELL, B.A.Sc.,   Edmonton, Alta.
City Engineer.

1.*H. J. McAUSLAN, B.A.Sc., O.L.S.,   North Bay, Ont.
Staff of T. & N.O. Ry.

3. J. A. McFARLANE. B.A.Sc,.   Hamilton, Ont.
Chief Draftsman, Hamilton Bridge Works Co.

*Diploma with honours.

1903—*Continued.*

1.*A. L. McNaughton,   Prince Rupert, B. C.
   With G.P.T. Co.

5.*F. G. Marriott, B.A.Sc.,   Toronto, Ont.
   Chemist and Supt. Asphalt Plant, City Testing Laboratory.

3.*C. A. Maus,   Paris, Ont.

3.*M. L. Miller,   Pittsburg, Pa.
   Draftsman, McClintic-Marshall Construction Co.

3. P. H. Mitchell,   Toronto, Ont.
   Consulting Electrical Engineer, Trader's Bank Building.

2.*R. H. Montgomery, B.A.Sc., O. and D.L.S.,   Prince Albert, Sask.
   Engineer and Surveyor.

1. F. A. Moore,   Winnipeg, Man.
   Engineering Dept., C. N. Ry.

3. E. E. Mullins,   Limon, Costa Rica, C.A.
   Baldwin Locomotive Works.

3. I. H. Nevitt, B.A.Sc.,   Toronto, Ont.
   Construction Bell Telephone Co.

1. E. W. Oliver, B.A.Sc., C.E.,   Toronto, Ont.
   Assistant Chief Engineer, Canadian Northern Ry. System.

3. J. P. Oliver,   Arabi, La.
   Supt. of Construction, The American Sugar Refining Co.

3. J. D. Pace, B.A.Sc.,   Montreal, Que.
   Construction Engineer, Canadian Westinghouse Co.

3. B. B. Patten, B.A.Sc.,   St. Catharines, Ont.
   Rutherford & Patten, Engineers and Surveyors.

2. D. H. Philp,   Ottawa, Ont.
   Georgian Bay Canal Survey.

3.*D. H. Pinkney,   Elvria, O.
   National Tube Dept., U.S. Steel Corporation.

2. T. H. Plunkett, B.A.Sc.,   Toronto, Ont.

1. D. F. Robertson, D.L.S.,   Ottawa, Ont.

3.*H. M. Schiebe, B.A.Sc.,   East Pittsburgh, Pa,
   Engineer, Westinghouse Electric & Mfg. Co.

1.*H. L. Seymour,   Edmonton, Alta.
   Sanders & Seymour. Civil Engineers and Dominion Land Surveyors.

1. J. H. Smith, D. & O.L.S.,   Edmonton, Alta.
   Engineer and Surveyor, 140 Jasper Ave.W.

3. H. G. Smith, B.A.Sc., (deceased).

3. S. L. Trees, B.A.Sc.,   Toronto, Ont.
   Supt. Mfg. Dept., Samuel Trees & Co., 42 Wellington St. East.

2. J. E. Umbach,   Ottawa, Ont.
   Topographical Surveys Branch, Department of the Interior.

1. J. Waldron, D.L.S.,   Moose Jaw, Sask.
   Engineer and Surveyor.

3.*S. B. Wass,   Presque Isle, Me.
   Chief Engineer, Amostook Valley Railroad.

3. J. A. Whelihan,   Regina, Sask.

*Diploma with honours.

1903—*Continued.*

3. H. F. WHITE,   London, Ont.
  Assistant Superintendent, The Geo. White & Sons Co., Ltd.
2.*C. G. WILLIAMS, B.A.Sc.,   Elk Lake, Ont.
  Superintendent. Otisse Mining Co.
1.*N. D. WILSON, B.A.Sc.,   Lumsden Bldg., Toronto.
  Engineer and Surveyor.
1.*C. R. YOUNG, B.A.Sc., A.M. Can. Soc. C.E.,   Toronto, Ont.
  Lecturer in Structural Engineering, University of Toronto.

1904.

3.*J. H. ALEXANDER, B.A.,   St. Louis, Mo.
  Engineer, Hunkins-Willis L. & C. Co.
3.*J. H. BARRETT,   Toronto, Ont.
  With the Wm. Davies Co., Ltd.
3. M. B. BONNELL,   Bobcaygeon, Ont.
3. T. D. BROWN, B.A.Sc.,   Calgary, Alta.
  Canadian Fairbanks Co.
3. F. W. BURNHAM, B.A.Sc.,   Milwaukee, Wis.
  Steam & Electrical Department, Allis-Chalmers-Bullock Co.
3. J. W. CALDER, B.A.Sc.,   Fort William, Ont.
  With Hydro-Electric Commission.
1. N. C. CAMERON,   4172 Dorchester St., Montreal, Que.
  Dominion Engineering and Construction Co.
1. A. J. CAMPBELL, B.A.Sc.,   Toronto, Ont.
3.*A. M. CAMPBELL, B.A.Sc.,   1403 King St. W., Toronto, Ont.
4. J. B. CHALLIES,   Ottawa, Ont.
  Hydraulic Engineer, Department of the Interior.
2. C. A. CHILVER   Walkerton, Ont.
2. H. L. CHILVER,   Moosehorn Bay, Man.
1. U. W. CHRISTIE, B.A.Sc., O.L.S.,   Ottawa, Ont.
  Astronomical Surveys Branch, Dept. of the Interior.
2. P. C. COATES, B.A.Sc.,   Cobalt, Ont.
  Mining Engineer.
1. S. B. CODE,   Smith's Falls, Ont.
  Town Engineer.
1.*T. F. CODE, B.A.Sc (deceased).
1.*W. A. COWAN,   Farnham, Que.
  Resident Engineer, C.P.R.
3.*S. F. CRAIG,   Toronto, Ont.
  Post-Graduate Course in Engineering, University of Toronto.
1.*S. R. CRERAR, B.A.Sc., O.L.S.,   Toronto, Ont.
  Lecturer in Surveying, University of Toronto.
3. W. M. CURRIE,   Hamilton, Ont.
  Managing Director, Canada Steel Co. Ltd.
3. H. H. DEPEW,   Fernie, B.C.
  Supt., Crow's Nest Pass Electric Light and Power Co.

*Diploma with honours.

1904—*Continued.*

2. A. J. ELDER, Ottawa, Ont.
Topographical Surveys Branch, Department of the Interior.

2. J. G. FLECK, Vancouver, B.C.
Fleck Bros. Ltd.

1.*A. L. FORD, B.A.Sc., Prince Rupert, B.C.
Government Inspector on G.T.P. Ry.

3. W. S. GIBSON, B.A.Sc., 38 Park Rd., Toronto, Ont.

1. J. N. GOODALL, Toronto, Ont.

1. J. P. GORDON, Toronto, Ont.
Engineering Staff, Willis Chipman, C.E.

3. W. W. GRAY, B.A.Sc., Toronto. Ont.
Lecturer in Mechanical Engineering, University of Toronto.

1. A. GRAY, B.A.Sc., Port Credit, Ont.
With St. Lawrence Starch Co.,

3. W. K. GREENWOOD, B.A.Sc., Orillia, Ont.
Town Engineer.

1. L. D. HARA, St. Catharines, Ont.
Assistant Engineer, Welland Canal Co.

3. C. J. HARRIS, B.A.Sc., Brantford, Ont.
With Brantford Screw Co.

1. J. B. HERON, B.A.Sc., Toronto, Ont.
c. o. S. H. Sykes, 1577 Danforth Ave.

1. E. M. M. HILL, Winnipeg, Man.
Engineering Dept., Canadian Northern Railway.

2. S. N. HILL, 325, Waverly St., Ottawa, Ont.
Topographical Surveys Branch, Department of the Interior.

2. C. J. INGLES, Niagara Falls, Ont.
With Ontario Power Co.

1. E. A. JAMES, B.A.Sc., Toronto, Ont.
Managing Editor, Canadian Engineer.

1. P. V. JERMYN, B. A.Sc., 118 King St. West, Toronto, Ont.
C.P.R. Construction Department.

3. W. S. H. KEEFE, Fort Covington, N.Y.
Manager, Light, Heat and Power Co.

3. W. J. LARKWORTHY (deceased).

3. O. B. McCUAIG, B.A.Sc., Wenatchee, Wash.
Supt. ,Entiat Light and Power Co.

1. G. G. McEWEN, B.A.Sc., Winchester, Ont.
Office of T. H. Dunn, O.L.S.

1.*W. G. McFARLANE, B.A.,B.A.Sc.,
Engineer and Surveyor, Peace River Dist.

3.*C. P. McGIBBON, B.A., East Pittsburgh, Pa.
With Westinghouse Electric and Mfg. Co.

3. C. McKAY, B.A.Sc., (deceased).

1. D. McMILLAN, Woodville, Ont

3. G. J. MANSON, Thorold, Ont.
With Manson Mfg. Co., Ltd.

*Diploma with honours.

1904—*Continued.*

1.*W. N. Moorhouse,   Toronto, Ont.
Office of Sproatt & Rolph, Architects.

3. E. E. Moore,   Glen Falls, N.Y·
Engineer, Inter-State Iron Co.

3. W. H. Munro,   Ottawa, Ont·
Assistant to J. B. McRae.

3. G. Pace, B.A.Sc.,   Hamilton, Ont·
With Canadian Westinghouse Co.

3. W. S. Pardoe, B.A.Sc.,   Philadelphia, Pa·
Instructor in Hydraulics, University of Pennsylvania.

3. J. Paris,   La Tuque, Que·
Resident Engineer, Trans. Ry.

2. J. Parke, B.A.Sc.,   Havilah, Ont.
Chemist and Assayer.

3. W. J. Parker,   Ottawa, Ont.
Top. Surveys Branch, Dept. of Interior.

3.*A. E. Pickering,   Sault Ste. Marie, Ont.
Supt., Lake Superior Power Co.

1. D. L. C. Raymond, B.A.Sc.,   Toronto, Ont.
Manager, The Concrete Engineering and Construction Co., Ltd.

1. F. B. Reid, B.A.Sc.,   Ottawa, Ont.
Astronomical Surveys Branch, Dept. of the Interior.

3.*M. R. Riddell, B.A.Sc.,   Toronto, Ont.

3. G. S. Roxburgh, B.A.Sc.,   Winnipeg, Man·
Manager, Fetherstonhaugh & Co., Patent Solicitors and Engineers.

2. F. N. Rutherford, B.A.Sc.,   St. Catharines, Ont.
Rutherford and Patten, Surveyors and Engineers.

1.*J. D. Sheply, B.A.Sc., D.L.S.,   N. Battleford, Sask.
District Surveyor and Engineer.

3. F. W. Slater, B.A.Sc.,   Schenectady, N.Y.
With General Electric Co.

3.*R. S. Smart,   Ottawa, Ont.
Manager, Fetherstonhaugh & Co., Patent Solicitors and Engineers.

1. D. A. Smith, B.A.Sc.,   Claude, Ont.

3. W. J. Smither, B.A.Sc.,   50 St. Clair Ave., Toronto, Ont.

3. S.E. Thomson, B.A.Sc.,   Niagara Falls, Ont.
Engineering Staff, Electrical Development Co.

3. C. J. Townsend, B.A.Sc.,   Toronto, Ont.

1. D. T. Townsend, B.A.Sc., O.L.S.,   Winnipeg, Man.
C.P.R. Land Department.

1. A. V. Trimble, B.A.Sc.,   Toronto, Ont.
Hydro-Electric Power Commission.

3. B. B. Tucker, B.A.Sc.,   Morrisburg, Ont
Resident Engineer, New York and Ontario Power Co.

2.*E. Wade, B.A.,   Welland, Ont.
Teacher.

1.*E. W. Walker, B.A.Sc., (deceased).

*Diploma with honours.

1904—*Continued.*

3. J. P. WATSON, B.A.Sc., .Montreal, Que.
   Draftsman, Motive Power Dept., C.P.Ry.
1. J. M. WEIR, Hamilton, Ont.
   Engineering Staff, G.T. Ry.
1.*A. F. WELLS, O.L.S.,B.A.Sc., Toronto, Ont.
   Wells & Gray, Ltd., Engineers and Contractors.
1. W. R. WORTHINGTON, B.A.Sc., Toronto, Ont.
   Assistant Sewer Engineer, Staff of City Engineer.
3. W. F. WRIGHT, Toronto, Ont.
   Sales Dept. Canadian General Electric Co.

1905.

2. H. W. ARENS, (deceased).
3. R. H. ARMOUR, 165 Broadway, New York.
   Westinghouse Electric & Manufacturing Co.
3.*C. B. AYLESWORTH, Hamilton. Ont.
   Draftsman, Canadian Westinghouse Co.
1.*W. BARBER, B.A.Sc., Toronto, Ont.
   Roadways Department, City Hall.
2.*W. A. BEGG, B.A.Sc.. Regina, Sask.
   Department of Public Works.
3.*G. G. BELL, Walkerville, Ont.
   Canadian Bridge Co.
1. J. C. BOECKH, Toronto, Ont.
   With Boeckh Brush Co.
3. W. M. BRISTOL, Halifax, N.S.
   Canadian Westinghouse Co.
2. W. C. CAMPBELL, Keene, Ont.
   Mining Engineer.
3. W. R. CARSON, High Bridge, N.J.
   Construction Engineer, Taylor Iron and Steel Co.
1. A. V. CHASE, Orillia, Ont.
3. S. R. A. CLEMENT, Toronto, Ont.
   With Hydro-Electric Power Commission.
3. T. E. CORRIGAN, Bodie, Cal.
   Chief Electrician, Standard Consolidated Mining Co.
1.*N. L. R. CROSBY. B.A.Sc., Chicago, Ill.
   Assistant Contracting Engineer, McClintic-Marshall Const. Co.
1. G. H. FERGUSON, B.A.Sc., Toronto, Ont.
   Hydro-Electric Commission.
3. H. S. FIERHELLER, B.A.Sc. (deceased).
3. F. W. HARRISON, 360 Pearl St., Brooklyn, N.Y.
   Chief Mechanical Draftsman, Edison Electric Illuminating Co.
1. M. C. HENDRY, B.A.Sc., Toronto, Ont.
2. C. S. L. HERTZBERG, Walkerville, Ont.
   Trussed Concrete Steel Co.

---

*Diploma with honours.

1905—*Continued.*

3. W. G. HEWSON, B.A.Sc.,   Toronto, Ont.
   With Smith, Kerry and Chace.
1. G. S. JONES,   Smith's Falls, Ont.
3. *G. KRIBS,   Toronto, Ont.
   With Smith, Kerry and Chace.
2. P. A. LAING,   Fanquier, On·.
   Resident Engineer, T.C. Ry.
1. A. LATORNELL, B.A.Sc.,   Toronto, Ont.
   Sewer Department, City Hall.
3. J. W. LEIGHTON,   Toronto, Ont.
   Secretary, Evans Rotary Engine Co.
1. *T. R. LOUDON, B.A.Sc.,   Toronto, Ont.
   Lecturer in Metallurgy, University of Toronto.
3. S. E· MCGORMAN,   Walkerville, Ont.
   Draftsman, Canadian Bridge Co.
1. *W. W. MCGREGOR (deceased).
2. D. W. MCKENZIE,   Winnipeg, Man.
   Draftsman, Engineering Dept., C.N. Ry.
3. *C. A. MCLEAN,   Toronto, Ont.
   Canadian Westinghouse Co.
2. W. N. MCLEAN.   Erin, Ont
3. F. G. MACE,   Ottawa, Ont.
   Patent Examiner, Dept. of Agriculture.
3. R. W. MOFFATT, B.A.Sc.,   Toronto, Ont.
   Demonstrator in Drawing, University of Toronto.
3. L. W. MORDEN,   Toronto, Ont.
   Canadian Westinghouse Co.
3. G. R. MUNRO, B.A.Sc.,   Peterboro', Ont.
   With Hudson Bay Survey.
3. *W. G. NICKLIN, B.A.Sc.,   Front & Tetellier Sts., Grand Rapids, Mich.
   Assistant Superintendent, Dalnu & Kiefer Tanning Co.
1. E. D. O'BRIEN,   Nipigon, Ont.
   With Transcontinental Ry.
1. *B. B. PATTEN, B.A.Sc.,   . St. Catharines, Ont.
   Rutherford and Patten, Surveyors and Engineers.
1. E. P. A. PHILLIPS, B.A.Sc., O.L.S.,   Porcupine, Ont.
   Pierce & Phillips, Engineers & Surveyors.
1. W. B. PORTE,   Oakville, Ont.
2. E. F. PULLEN,   Cochrane, Ont·
   Resident Engineer, Transcontinental R.R.
2. G. L. RAMSEY, B.A.Sc.,   Dunnville, Ont.
1. G. W. RAYNER,   Thorold, Ont.
3. *R. B. ROSS, (deceased).
5. T. E. ROTHWELL, B.A.Sc.,   Belleville, Ont.
   Provincial Assay Office.
2. *G. S. SCOTT,   Toronto, Ont.
   Fellow in Mineralogy, University of Toronto.

*Diploma with honours.

1905—*Continued.*

3. H. V. SERSON, Highbridge, N.J.
Engineer in charge, Power House Const., Taylor Iron & Steel Co.

3. C. H. SHIRRIFF, B.A.Sc., Toronto, Ont.
Chemist, Imperial Extract Company.

3.*C. E. SISSON, Peterboro', Ont.
Engineering Department, Canadian Gen. Electric Co.

1. D. L. N. STEWART, B.A.Sc., Collingwood, Ont.

1. M. A. STEWART, Toronto, Ont.
Assistant City Engineer, Roadway Dept., City Hall.

3.*W. F. STUBBS, Galt, Ont.
Assistant Engineer, Goldie & McCulloch Co.

1. N. H. STURDY, Cleveland, Ohio.
Designer, L, S. & M. S. Ry.

1. W. G. SWAN, B.A.Sc., Langley, B.C.
Divisional Engineer, C. N. Ry.

1.*F. H. SYKES, O.L.S., D.L.S., Toronto, Ont.
Asst. Structural Engineer, with City Architect, City Hall.

3. L. R. THOMSON, B.A.Sc., Winnipeg, Man.
Lecturer in Civil Engineering, University of Manitoba.

3. E. D. TILLSON, B.A.Sc., New York, N.Y.
Engineer for Const. Dept., Safety Insulated Wire & Cable Co.

1.*J. J. TRAILL, B.A.Sc., Toronto, Ont.
Lecturer in Mechanical Engineering. University of Toronto.

1.*W. M. TREADGOLD, B.A., Toronto, Ont.
Lecturer in Surveying, University of Toronto.

3. W. E. TURNER, B.A.Sc., Salt Lake City, Utah.
With Utah Light & Ry. Co.

3. A. E. UREN, Toronto, Ont.
Editor, Acton Publishing Co.

3. J. M. VAUGHAN, 58 Melville Ave., Toronto, Ont.
Contractor.

1. H. L. WAGNER, B.A.Sc., Hamilton, Ont.
Draftsman, Hamilton Bridge Works Co.

2. W. H. YOUNG, B.A.S c.,D.L.S., Lethbridge, Alta.
Hamilton & Young, Dominion Land Surveyors and Engineers.

1906.

1. F. ALPORT, Superior Junct., Ont.
Resident Engineer, Transcontinental Ry.

3.*W. L. AMOS, Guelph. Ont.

1. A. H. ARENS, Inverness, N.S.
Resident Engineer. Inverness Ry. & Coal Co.

3.*J. C. ARMER, B.A.Sc., Toronto, Ont.
Secretary and Managing Editor of the Canadian Manufacturer Pub. Co. Ltd.

1. M. H. BAKER, B.A.Sc., St. Thomas, Ont.
Assistant City Engineer.

*Diploma with honours.

1906—*Continued.*

3. F. W. BALDWIN,   Baddeck, N.S.
With Graham Bell, Esq.

2. E. W. BANTING, B. A.Sc.,   Toronto, Ont.
Demonstrator in Surveying, University of Toronto.

3. F. BARBER, B.A.,   Toronto, Ont.
York County Engineer, 57 Adelaide St. E.

2. M. BATES, B.A.Sc (deceased).

2. J. P. BELLISLE, (deceased).

3.*H. H. BETTS, B.A.Sc.,   Rio de Janiero, Brazil.
Rio de Janiero Tramway, Light & Power Co.

5.*D. E. BEYNON, B.A.Sc.,   Toronto, Ont.
With Dunlop Rubber Goods Co.

2. G. W. BISSETT,   Naughton, Ont.
Mill Supt., Canadian Exploration Co., Ltd.

3. W. C. BLACKWOOD, B.A.Sc.,   Toronto, Ont.
Demonstrator in Physics, University of Toronto.

3. H. E. BRANDON, B.A.Sc.,   Winnipeg, Man.
Chief Engineer, Vulcan Iron Works.

1. M. E. BRIAN, B.A.Sc., O.I.S., A.M. Can. Soc. C.E.,   Windsor, Ont.
City Engineer.

2. T. W. BROWN, B.A.Sc., D. & S.L.S.,   Saskatoon, Sask.
Civil Engineer.

1.*A. E. K. BUNNELL, B.A.Sc.,   Weyburn, Sask.
Engineering Staff, Willis Chipman, C.E.

3. F. M. BYAM,   Toronto, Ont.
With Smith, Kerry, and Chace.

3. A. CAMERON,   Toronto, Ont
Draftsman, Canada Foundry Co.

3. A. W. CAMPBELL, B.A.Sc.,   Toronto, Ont.
Inspector, Hydro-Electric Power Commission.

1. M. J. CARROLL,   Ottawa, Ont.
Topographical Surveys Branch, Department of the Interior.

3.*R. E. C. CHADWICK,   Toronto, Ont.
Staff of City Engineer.

1.*G. T. CLARK, B.A.,   Saskatoon, Sask.
City Engineer.

3.*G. A. COLHOUN,   Hamilton, Ont.
Draftsman, The Hamilton Bridge Works Co., Ltd.

1.*W. A. M. COOK, B.A.Sc.,   Toronto, Ont.
Staff of City Architect, City Hall.

1.*E. L. COUSINS, B.A.Sc.,   Toronto, Ont.
Resident Engineer, G. T. Ry., Middle and Southern Div.

4. A. G. CREIGHTON,   Prince Albert, Sask.
Creighton & McConnell, Architects and Structural Engineers.

4. W. N. DANIELS,   1215 Filbert St., Philadelphia, Pa.
With John R. Wiggins & Co.

3.*N. P. F. DEATH, B.A.Sc.   25 Jarvis St., Toronto, Ont.
Death & Watson, Electrical Engineers and Contractors.

---

*Diploma with honours.

1906—*Continued.*

3. C. S. Dundass, B.A.Sc.   Lachine, Que.
   With Dominion Bridge Co.
3. S. L. Fear,   Amherstburg, Ont.
   With Dunbar, Sullivan Dredging Co.
5.*C. C. Forward,   Ottawa, Ont.
   Laboratory of the Inland Revenue Department.
5. C. W. Graham, B.A.Sc.,   Toronto, Ont.
   Industrial Chemist, Wm. Davies Co.
1.*P. W. Greene.   Toronto, Ont.
   With Hydro-Electric Power Commission.
3. C. B. Hamilton, B.A.Sc.,   43 Madison Ave., Toronto, Ont.
   Hamilton Gear and Machinery Co.
1.*A. L. Harkness, B.A.Sc.,   Lachine Locks, Que.
   With Dominion Bridge Co., Ltd.
1.*R. L. Harrison,   Cobourg, Ont.
   Resident Engineer, Canadian Northern Ry.
1. E. Harrison, B.A.Sc.,   New Liskeard, Ont.
   With Sutcliffe & Neelands.
3. J. C. Hartney, B.A.Sc.,   Vancouver, B.C.
   Engineer & Salesman. Canadian Westinghouse Co.
1. S. Hett, B.A.Sc.,   Sutton West, Ont.
   Surveyor.
3. C. R. Hillis,   Toronto, Ont.
   With Toronto & Niagara Power Co.
3. C. W. Hookway, B.A.Sc.,   Winnipeg, Man.
   Allis-Chalmers Bullock Co.
3. R. H. Hopkins, B.A.Sc.,   Toronto, Ont.
   Demonstrator in Electrical Engineering, University of Toronto.
1.*R. S. Houston,   Emerson, Man.
2.*W. Huber,   Toronto, Ont.
   With Canadian Inspection Co.
3.*A. H. Hull, B.A.Sc.,   Toronto, Ont.
   With Smith, Kerry and Chace.
3. W. C. Jepson,   Niagara Falls, Ont.
   Welland Canal Office.
1.*C. Johnston, B.A.Sc.,   Toronto, Ont.
   District Engineer, Can. Northern Ry.
1. G. R. Jones, B.A.Sc.,   China.
   Missionary.
3. T. Jones, B.A.Sc.,   18 Meredith Crescent, Toronto, Ont.
1.*A. E. Jupp, B.A.Sc.,   Haileybury, Ont.
3. J. D. Keppy,   Toronto, Ont.
   Canadian Inspection Co.
1. J. L. Lang, B.A.Sc., D. & O.L.S.,   Sault Ste. Marie, Ont.
   Lang & Keys, Engineers and Surveyors.
3. A. P. Linton, B.A.Sc.,   Montreal Que.
   With Dominion Bridge Co.

---

*Diploma with honours.

1906—*Continued.*

4.*A. WELLESLEY MCCONNELL, B.A.Sc.,   Toronto, Ont.
Lecturer in Architecture, University of Toronto.

3.*D. G. MCILWRAITH,   Galt, Ont.
Draftsman, The Goldie & McCulloch Co., Ltd.

2. J. A. MCKENZIE,   Cobalt, Ont.
Manager, Nipissing Reduction Co.

1.*J. V. MCNAB,   Kenora, Ont.
Transitman, C.PR. Engineering Staff.

3. J. A. MCPHERSON,   Toronto, Ont.
Student, Faculty of Medicine, University of Toronto.

2. K. A. MACKENIZE, B.A.Sc.,   Toronto, Ont.
Managing Editor, "Builder & Contractor."

1. W. MacKinnon,   Rankin, Pa.
Erection Dept., McClintic Marshall Construction Co.

3.*W. MACLACHLAN, B.A.Sc.,   Belleville. Ont.
Trenton Electric and Waterpower Co.

3.*D. W. MARRS,   Pittsburg, Pa
Designer and Estimator, Riter-Conley Mfg., Co.

3. W. A. MAXWELL,   Walkerville, Ont.
Draftsman, Canadian Bridge Co.

1.*REV. J. MELLON MENZIES, B.A.Sc., D.L.S.,   North Honan, China.
Missionary, Wu An

3. L. R. MILLER, B.A.Sc.,   Orillia, Ont.

1.*B. F. MITCHELL, B.A.Sc.,   Edmonton, Alta.
Municipal Engineer.

1. F. F. MONTAGUE,   506 Union Bank Bldg., Winnipeg, Man.
Law Student.

1.*W. J. MOORE, O.L.S.,   Pembroke, Ont.
Morris and Moore, Land Surveyors and Architects.

1. C. R. MURDOCK,   Weyburn, Sask.

2. C. J. MURPHY, B.A.Sc.,   Fernie, B.C.
Assistant Engineer, Crow's Nest Pass Coal Co.

1.*W. P. NEAR, B.A., B.A.Sc.,   Toronto, Ont.
Staff of City Engineer.

2. R. NEELANDS,   Port Hammond, B.C.

3. D. G. PARK, B.A.Sc.,   West Allis, Wis.
With Allis-Chambers-Bullock Co.

3. G. W. PATERSON,   Vancouver, B.C.
Salesman, Canadian Financiers, Ltd.

5. R. E. PETTINGILL,   Port Colborne, Ont.
Chief Chemist, Portland Cenemt Co.

2.*R. C. PURSER, B.A.Sc.,   Windsor, Ont.

3. N. R. ROBERTSON, B.A.Sc.,   Walkerton, Ont.

1. J. O. RODDICK, B.A.Sc.,   Toronto, Ont.
Assistant Engineer, Dept. of Public Works of Canada.

1. C. H. ROGERS, B.A.Sc.,   Peterboro' Ont.
Peterboro Canoe Co.

---

*Diploma with honours.

1906—*Continued.*

2.*O ROLFSON, B.A.Sc., D.L.S..   Peace Kiver District.
Dominion Land Surveyor.

1. R. C. ROSS, B.A.Sc.,   Ottawa, Ont.
Department of the Interior.

1. K. G. ROSS,   Sault Ste. Marie, Ont.
With Lang & Keys, Engineers and Surveyors.

1.*H. T. ROUTLY, O.L.S., D.L.S ,   Haileybury, Ont.
Routly, Summers & Malcolmson, Engineers. amd Surveyors.

2. J. H. RYCKMAN,   Chicago, Ill.
Bridge and Bldg. Dept., Chicago Milwaukee & St. Paul Ry.

3.*W. K. SANDERS.   58 Webster St., West Newton, Mass.

1.*W. A. SCOTT, B.A.Sc., D.L.S.,   Galt, Ont.
Dominion Land Surveyor.

1.*W. M. STEWART, B.A.Sc.,   142 Aberdeen Ave., Hamilton, Ont.

2. J. E. THOMSON, B.A.Sc.,   W. Virginia, U. S. A.
With Sterling Coal Co.

3.*C. L. VICKERY,   112 Barlow St., Fall River, Mass.
Chief Engineer, American Thread Co.

5. W. E. WICKETT (deceased).

3.*J. N. WILSON, B.A.Sc.,   Toronto, Ont.
Electrical Dept., City of Toronto.

3*E. M. WOOD, B.A.Sc.,   136 Lee Ave., Toronto, Ont.

1907.

3.*F. G. ALLEN, B.A.Sc.,   Hyde Park, Mass.
Assistant to Chief Engineer, B.F. Sturtevant Co.,

1. F. J. ANDERSON, B.A.Sc.,   Niagara Falls, Ont.

1. A. P. AUGUSTINE,   Vancouver, B.C.
B. C. Land Surveyor.

3.*H. D. BOWMAN, B.A.Sc.,   Niagara Falls, Ont.
With the Ontario Power Co.

3. W. S. BRADY, B.A.Sc.,   Toronto, Ont.

1. G. H. BROUGHTON,   Penticton, B.C.

1. J. A. BROWN, B.A.Sc.,   Toronto, Ont.
Fellow in Drawing, University of Toronto.

1. C. E. BUSH, B.A.Sc.,   285 College St., Toronto, Ont.

3. J. H. CASTER,   Peterboro', Ont.
Production Dept. Can. Gen. Elec. Co.

1.*E. CAVELL,   Saskatoon, Sask.
Wiggins & Cavell, Surveyors and Engineers.

1.*C. B. B. CONNELL,   Glasgow, Scotland.
With Mirrless & Watson.

3.*G. C. COWPER, B.A.Sc.,   Frankford, Ont.
Trent Valley Canal.

2. J. V. CULBERT, B.A.Sc.,   Toronto, Ont.
Staff of Willis Chipman, C.E.

---

*Diploma with honours.

1907—*Continued.*

3.*R. S. DAVIS, B.A.Ss.,   Calgary, Alta.
 Sales Engineer, Canadian Westinghouse Co.

3. S. D. EVANS, B.A.Sc.,   Leamington, Ont.

3.*F. R. EWART, B.A.Sc.,   Toronto, Ont.

1. G. R. S. FLEMING,   Toronto, Ont.
 With Atwell Fleming Printing Co.

6. P. C. FUX, B.A.Sc.,   Brantford, Ont.
 With Waterous Engine Works Co.

1. J. S. GALLETLY,   Toronto, Ont.
 Post-Graduate Course in Engineering, University of Toronto.

2. G. GALT, B.A.Sc.,   East Ely, Nevada.
 With Nevada Co.

1. A. B. GARROW, B.A.Sc.,   Toronto, Ont.
 Staff of City Engineer.

1. A. GILLIES, B.A.Sc.,   Fort William, Ont.

1. G. W. GRAHAM,   Eugenia, Ont.

3. C. S. GRASETT, B.A.Sc.,   8 Harbord St., Toronto, Ont.

1.*R. E. W. HAGARTY, B.A.Sc.,   Toronto, Ont.
 Demonstrator in Drawing, University of Toronto.

3. K. HALL, B.A.Sc.,   39 Sherman Ave., Hamilton.

1. C. T. HAMILTON, B.A.Sc.,   Niagara Falls, Ont.
 With Ontario Power Co.

3. R. A. HARE.   St. Catharines, Ont.

3.*H. O. HILL, B.A.Sc.,   Pittsburgh, Pa.
 With Riter-Conley Mfg. Co.

1.*T. H. HOGG, B.A.Sc.,   Niagara Falls, Ont.
 Assistant to Supt. of Construction, Ontario Power Co.

3.*C. H. HUTTON, B.A.Sc.,   Hamilton, Ont.
 Engineering Staff, Dominion Power Co.

1. H. M. HYLAND, B.A.Sc., 72 St. Mary St., Toronto, Ont.

3. E. W. HYMAN, B.A.Sc.,   London, Ont.
 Assistant Superintendent, London Electric Co.

3.*L. G. IRELAND, B.A.Sc.,   Belleville, Ont.
 Supt., Midland Construction Co., Ltd.

1.*W. JACKSON, B.A.Sc.,   Niagara Falls, Ont.
 With Ontario Power Co.

4.*C. B. JACKSON,   Chicago, Ill.
 Estimating Dept. E. Everett Clora Co.

3.*E. W. KAY, B.A.Sc.,   Winnipeg, Man.
 Salesman, Canadian Westinghouse Co.

3. D. F. KEITH,   Provo, Utah.
 Electrical Engineer, Telluride Power Co.

1. H. P. KEITH   Comber, Ont.

1. A. A. KINGHORN, B.A.Sc.,   Toronto, Ont.
 Inspector of Road ways,City Engineers Department.

 *Diploma with honours.

1907—*Continued*.

1. L. W. KLINGNER 568 Spadina Ave., Toronto, Ont.
1.*F. C. LAMB, B.A.Sc., North Battleford, Sask.
 Department of Public Works.
3. A. D. LePAN, B.A.Sc., Toronto, Ont.
 Assistant Superintendent of Buildings and Grounds, University of Toronto.
1. J. H. LINDSAY. Toronto, Ont.
 Fellow in Surveying, University of Toronto.
3. J. A. D. McCURDY, Hammondsport, N.Y.
 With Graham Bell, Esq.
1.*J. B. McFARLANE, B.A.Sc., 60 Lonsdale Rd., Toronto, Ont.
 Dominion Land Surveyor.
3.*D. J. McGUGAN, B.A.Sc., Vancouver, B.C.
 With Humphreys & Tupper, Civil Engineers and Surveyors.
3. A. H. McINTOSH, Chicago, Ill.
 With Illinois Steel Co.
3. F. W. McNEILL, B.A.Sc., Peterboro', Ont.
 Canadian General Electric Co.
1.*M. K .McQUARRIE, Vancouver, B.C.
 Terminal Engineer, C.P. Ry. Co.
1.*G. MacLEOD, Montreal, P.Q.
 Assistant Secretary, Can. Soc. C.E.
1. A. G. MACKAY, New York, N.Y.
 With Hudson & Manhattan Ry. Co.
1. W. S. MALCOLMSON, B.A.Sc., Haileybury, Ont.
 Roully, Summers & Malcolmson, Engineers and Surveyors.
3. S. A. MARSHALL, Snelgrove, Ont.
6. D. H. C. MASON, B.A.Sc., Toronto, Ont.
1. J. W. MELSON, B.A.Sc., Toronto, Ont.
 Fellow in Drawing, University of Toronto.
1. G. G. MILLS, B.A.Sc., Toronto, Ont.
3. J. B. MINNS, B.A.Sc., Peterboro', Ont.
 Canadian General Electric Co.
4.*G. N. MOLESWORTH, Toronto. Ont.
 Draftsman, Eden Smith & Son. Architects.
1. J. M. MOORE, B.A.Sc., London. Ont.
 With McClary Mfg. Co.
5.*P. F. MORLEY. 177 Pearson Ave., Toronto. Ont.
1. E. W. MURRAY, Stratford, Ont.
 Engineer, with Edge & Gutteridge, Contractors and Builders.
3. J. D. MURRAY, Toronto, Ont.
 With Fetherstonhaugh & Co., Patent Solicitors and Engineers.
1. E. W. NEELANDS, B.A.Sc., New Liskeard, Ont.
 Sutcliffe & Neelands, Consulting Engineers.
1. R. E. K. NEELANDS, B.A.Sc., Brampton Ont.
2.*B. NEILLY, B.A.Sc.. Cobalt, Ont.
 Assayer, Black Consolidated.

---

*Diploma with honours.

1907—*Continued.*

1. A. E. NOURSE, B.A.Sc.,   Toronto, Ont.
Assistant Engineer, Expanded Metal Co.

3. J. J. O'SULLIVAN,   Toronto, Ont.
With Canada Railway News Co.

2. T. K. PATON,   Wardner, Ida.
Mining Engineer.

1. F. W. PAULIN, O.L.S.,   Niagara Falls, Ont.
Civil Engineer.

1. R. B. POTTER,   Toronto, Ont.
Post-Graduate course in Engineering. University of Toronto.

3.*F. E. PROCHNOW, B.A.Sc.,   Buffalo, N.Y.
With Wilhelm, Parker & Ward, Patent Attorneys.

3.*J. F. PROCUNIER,   Bayham. Ont.

3. G. E. QUANCE, B.A.Sc.,   Delhi, Ont.

3.*H. RAINE,   Hamilton, Ont.
With Hamilton Bridge Works Co.

1.*J. L. RANNIE, B.A.Sc.,   Ottawa, Ont.
Observer, Geodetic Survey.

3. C. W. B. RICHARDSON, B.A.Sc.,   Montreal, Que
Motive Power Dept., Angus Shops.

1. A. A. RIDLER,   Toronto, Ont.
Supt., Constructing & Paving Co., Ltd.

5. H. E. ROTHWELL,   Port Richmond, N.Y.
Assistant Chemist, Standard Varnish Works.

5. C. A. SCHOFIELD,   Buffalo, N.Y.
Chemist, Schoell-Kopf-Hartford & Hanna Co.

1.*A. C. T. SHEPPARD,   Ottawa, Ont.
Department of Mines.

1. F. R. SMITH, B.A.   Gowganda, Ont.
Manager, Can. Gowganda Silver Mines.

3. E. R. SMITHRIM, B.A.Sc.,   Sault Ste., Marie, Ont.
With Tagona Water and Light Co.

1.*W. SNAITH,   Toronto, Ont.
Assistant Engineer, Barber & Young.

3. A. C. SPENCER, B.A.Sc.,   London, Ont.
Mechanical Engineer, McClary Mfg. Co.

3. G. S. STEWART,   Toronto, Ont.
Agent, Canadian General Electric Co.

1. J. A. STILES, B.A.Sc.,   Toronto, Ont.
Demonstrator in Drawing, University of Toronto.

3.*J. L. STIVER,   Ottawa, Ont.
Electrical Standard Laboratory, Inland Revenue Department.

1. J. L. G. STUART, B.A.Sc.,   Toronto, Ont.
Railway and Special Works Department, City Hall.

1. G. F. SUMMERS, O.L.S.,   Haileybury, Ont.
Routly, Summers & Malcolmson, Engineers and Surveyors.

*Diploma with honours.

1907—*Continued.*

1.*H. W. SUTCLIFFE, New Liskeard, Ont.
Sutcliffe & Neelands, Consulting Engineers.
1. P. M. THOMPSON, B.A.Sc., Ambridge. Pa.
Draftsman, American Bridge Co.
3. O. R. THOMSON, B.A.Sc., Berlin, Ont.
Inspector, Hydro-Electric Power Commission.
1. L. R. THOMSON, B.A.Sc., Winnipeg, Man.
Lecturer in Civil Engineering, University of Manitoba.
1. W. J. WALKER, Nipigon, Ont.
With Transcontinental Ry.
1. E. D. WILKES, B.A.Sc., Toronto, Ont.
Main Drainage Department, City Hall.
3. A. F. WILSON, B.A.Sc., Chicago, Ill.
Inspector, Ghicago, Telephone Co.
3. M. H. WOODS, Toronto, Ont.
Post-Graduate Course in Engineering, University of Totonto.
1. G. W. A. WRIGHT, 517 Oxford St., London, Ont.
Warren Bituminous Paving Co.
3.J. YOUNG, Toronto, Ont.
Inspector. Canadian Fire Underwriter's Association.
3.*A. R. ZIMMER, B.A.Sc.. Toronto, Ont.
Demonstrator in Electrical Engineering, University of Toronto.

1908.

3. H. G. AKERS, B.A.Sc., Toronto, Ont.
Demonstrator in Electrochemistry, University of Toronto.
3. L. F. ALLAN, Toronto, Ont.
Roadway Dept., City Hall.
1.*C. B. ALLISON, South Woodslee, Ont.
1.*R. M. ANDERSON, B.A.Sc.. Burlington, Ont.
5. J. R. ARENS, Toronto, Ont.
Post -graduate Course in Engineering, University of Toronto.
3. H. C. BARBER, Toronto, Ont.
Post-graduate Course in Engineering, University of Toronto.
1. E. BARTLETT, B.A.Sc., Smithville, Ont.
Dominion Land Survey Work.
2. F. J. BEDFORD, Porcupine, Ont.
1.*G. G. BELL, Portland, Me.
With Sawyer & Moulton, Consulting Engineers.
3. G. E. BLACK, B.A.Sc., Toronto, Ont.
Public Works Department of Ontario.
3. H. F. BOWES, Toronto, Ont.
Superintendent of Warren Bituminous Paving Co., Ltd.
3.*J. H. BRACE, Brooklyn. N.Y·
With N.Y. Telephone Co.

*Diploma with honours.

1908—*Continued.*

1. P. R. BRECKEN. B.A.Sc.,   Montreal, Que.
    General Secretary, Y.M.C.A.
3. E. I. BROWN,   Nipissing, Ont.
    Assistant Engineer on Const. Can. Westinghouse Co., Nipissing Power Company.
1. W. F. M. BRYCE,   Ottawa, Ont.
    Assistant Engineer. City Engineer's Department.
3. P. H. BUCHAN, B.A.Sc.,   Vancouver, B.C.
    Engineering Department, B.C. Electric Ry. Co., Ltd.
2. J. F. CAMPBELL, B.A.Sc.,   Copper Cliff, Ont.
    Canadian Copper Co.
3. N. A. CAMPBELL,   Calgary. Alta.
    Chief Chemist. Canada Cement Co.
3. A. M. CARROLL,   Cobalt, Ont.
    Manager, Rochester Cobalt Mines, Ltd.
1. H. R. CARSCALLEN, B.A.Sc.,   Calgary, Alta.
    Assistant Hydrographer, with P. M. Sauder.
3. G. CHALLEN,   Goderich, Ont.
    1. F. H. CHESNUT, B.A.Sc.,   Mount Lebanon, B.C.
    Resident Engineer, C.N.R. Plant.
1. W. E. COLE (deceased).
4.*W. C. COLLETT, B.A.Sc..   Toronto, Ont.
    Draftsman, F. S. Baker Trader's Bank Bldg.
1. R. Y. CORY. B.A.Sc.,   530 Huron St., Toronto, Ont.
3.*H. COYNE, B.A.Sc.,   5930 South Park Ave., Chicago. Ill.
    Designing Draftsman.
2.*J. D. CUMMING, B.A.Sc.,   Copper Cliff, Ont.
6. A. D. DAHL, B.A.Sc.,   Midland, Mich.
    Chemist. Dow Chemical Co.
1. F. A. DANKS,   Box 230. Penticton, B.C.
3. J. DARROCH,   Detroit, Mich.
    Draftsman, Autoparts Mfg. Co.
3. H. C. DOORLY,   6 McClellan St., Schenectady, N.Y.
2. R. H. DOUGLAS,   Edmonton, Alta.
    Department of Public Works,
2.*F. C. DYER, B.A.Sc.,   Toronto, Ont.
    Demonstrator in Mining. University of Toronto.
1. F. M. EAGLESON,   Gorrie, Ont.
1. C. Edwards, B.A.Sc.,   Toronto. Ont.
    With Standard Sanitary Mfg. Co.
1. S. L. EVANS. B.A.Sc.,   Corinth, Ont.
1. E. O. EWING,   Campbellford, Ont.
1. O. L. FLANAGAN, B.A.Sc.,   Toronto, Ont.
    With C. H. Mitchell, Consulting Engineer, 1003 Traders' Bank Bldg.
1. C. FLINT, B.A.Sc.,   Monadnock, B.L., Chicago.
    With MacDonald Engine Co.

*Diploma with honours.

1908—*Continued.*

1. A. H. FOSTER, B.A.Sc.,   Toronto, Ont.
3. G. C. FRANCIS,   Toronto, Ont.
    City Engineer's Staff, Roadway's Department.
3. S. S. GEAR.   Fort Erie.
1. C. A. GRASSIE, B.A.Sc.,   Welland, Ont.
3.*C. L. GULLEY, B.A.Sc.,   Toronto, Ont.
    Demonstrator in Electrical Engineering, University of Toronto.
3. J. W. HACKNER, B.A.Sc.,   Nairn Centre, Ont.
    Inspector of Public Works.
3. F. L. HAVILAND,   Hamilton, Ont.
    Draftsman, Hamilton Bridge Works Co.
1.*C. D. HENDERSON,   Walkerville, Ont.
    Canadian Bridge Co.
5.*D. J. HUETHER, B.A.Sc.,   Plainfield, N.J.
    Manager, Century Rubber Trading Co.
1. A. D. HUETHER, B.A.Sc.,   77 Grenville St., Toronto, Ont.
3.*A. N. HUNTER, B.A.Sc.,   Toronto, Ont.
    Demonstrator in Electrical Engineering, University of Toronto.
3. S. B. ILER,   Belleville, Ont.
    Construction Department, Seymour Power & Electric Co.
1.*J. T. JOHNSTON, B.A.Sc.,   Campbellford, Ont.
    Assistant Engineer on Construction of Trent Canal.
2. H. G. KENNEDY,   Toronto, Ont.
    Post Graduate Course in Engineering, University of Toronto.
1.*W. R. KEYS,   Winchester, Ont.
3.*J. N. M. LESLIE, B.A.Sc.,   Hamilton, Ont.
    With Canadian Westinghouse Co.
3. F. C. LEWIS,   Chicago, Ill.
    American Bridge Co.
3. H. R. LYNAR,   Toronto, Ont.
    Lynar & Mace, Consulting Engineers.
1.*W. G. McGEORGE,   Chatham, Ont.
    Consulting Engineer.
1. J. M. McGREGOR.   Ridgetown, Ont.
1. L. A. McLEAN, B.A.Sc., (deceased).
1. W. A. A. McMASTER,   Palmerston, Ont.
1. H. C. McMORDIE, B.A.Sc.,   Walkerville, Ont.
    With Canadian Bridge Co.
1.*A. A. McROBERTS, B.A.Sc.,   Pontypool, Ont.
    Assistant on Dominion Land Survey.
5.*N. G. MADGE,   New York, N. Y.
    Chief Chemist, Continental Rubber Co. of N. Y.
3. J. E. MALONE,   Chicago, Ill.
    With Illinois Steel Co.
5. K. D. MARLATT,   Oakville, Ont.
    The Marlatt & Armstrong Co.

*Diploma with honours.

1908—*Continued.*

1. R. J. MARSHALL, B.A.Sc.,   Toronto, Ont.
Demonstrator in Applied Mechanics, University of Toronto.

5. G. L. MILLIGAN,   Toronto, Ont.
Post-graduate Course in Engineering, University of Toronto.

1. A. B. MITCHELL,   Cambridge, Mass.
Student in Architecture, Harvard University.

4.*J. C. P. MOLESWORTH (deceased).

3. E. D. MONK, B.A.Sc.,   Pittsfield, Mass.
Testing Department, General Electric Co.

3.*F. H. MOODY, B.A.Sc.,   49-55 Lafayette St., New York, N.Y.
Associate Editor of "Machinery."

3. J. H. MORICE, B.A.Sc.,   938 Albany, St., Schenectady, N.Y.

3. F. E. H. MOWBRAY, B.A.Sc.,   Hamilton, Ont.
Canadian Westinghouse Co.

3.*W. P. MURRAY, B.A.Sc.,   Montreal, Que.
Dominion Bridge Co.

3. W. deC. O'GRADY,   Winnipeg, Man.
Engineer, Gas Traction Co., Ltd.

1. H. J. PECKOVER, B.A.Sc.,   Toronto, Ont.
Fellow in Drawing, University of Toronto.

1.*M. PEQUEGNAT, B.A.Sc.,   Toronto, Ont.
Demonstrator in Drawing, University of Toronto.

1. H. G. PHILLIPS,   Saskatoon, Sask.
Phillips & Phillips, Civil Engineers and Surveyors.

3. M. PIVNICK,   Toronto, Ont.
Student in Dentistry.

1.*E. M. PROCTOR, B.A.Sc.,   Toronto, Ont.
Draftsman, Canada Foundry Co.

3.*C. F. PUBLOW, B.A.Sc.,   305 Gladstone Ave., Toronto, Ont.

1. J. T. RANSOM, B. A.Sc.,   Toronto, Ont.
Fellow in Surveying, University of Toronto.

1.*W. B. REDFERN, B.A.Sc.,   Steelton, Ont.
Resident Engineer for Willis Chipman, C.E.

1. F. L. RICHARDSON, B.A.Sc.,   Toronto, Ont.
With Miller, Cummings & Robertson.

3. H. A. RICKER, B.A.Sc.,   101 Leeming St., Hamilton. Ont.

1. A. R. ROBERTSON, B.A.Sc.,   Toronto, Ont.
Staff of City Engineer.

5. F. A. ROBERTSON,   Toronto, Ont.
Assistant Advertising & Business Manager, Farmers' Magazine.

1.*W. A. ROBINSON,   Winnipeg, Man.

3. R. C. ROBINSON,   134 Edmonton St., Winnipeg, Ont.

5. L. J. ROGERS,   Toronto, Ont.
Demonstrator in Chemistry, University of Toronto.

2.*R. R. ROSE, B.A.Sc.,   Copper Cliff, Ont.
With Canadian Copper Co.

3. D. ROSS,   387 Princess Ave., London Ont.

*Diploma with honours.

1908—*Continued.*

1. A. O. SECORD, Brantford, Ont.

3. W. E. V. SHAW, B.A.Sc., Toronto, Ont.
Purchasing Engineer, Hydro-Electric Power Commission.

3. H. F. SHEARER, B.A.Sc., E. Norwood, Ohio.
Testing Engineer, Bullock Electric Mfg. Co.

1. W. L. STAMFORD, B.A.Sc.. Point du Bois, Man
Inspector on Concrete Work, Hydro-Electric Power Plant.

3. R. H. STARR, B.A.Sc., Toronto, Ont.
Toronto Hydro-Electric System, City Hall.

3. A. W. J. STEWART, Toronto, Ont.
Electrical Department, City Hall.

3. J. St. LAWRENCE, Erie, Pa.
Supt. of Engine Shops, Erie City Iron Works.

1. J. J. STOCK, Toronto, Ont

1. H. B. STUART, B.A.Sc., Hamilton, Ont.
Draftsman, Hamilton Bridge Works Co.

2. J. L. G. STUART, B.A.Sc., Toronto, Ont.
Railway & Special Works Department, City Hall.

3. A. D. SWORD, Toronto, Ont.
Post-Graduate Course in Engineering, University of Toronto.

3. J. W. R. TAYLOR, B.A.Sc., Hamilton, Ont.
With Canadian Westinghouse Co.

1.*W. E. TAYLOR, B.A.Sc., Toronto, Ont.
Fellow in Drawing, University of Toronto.

3. V. C. THOMAS, B.A.Sc., Toronto, Ont.
Demonstrator in Hydraulics, University of Toronto.

1. J. H. THORNLEY, B.A.Sc., 943 Dundas St., London. Ont.

1. C. G. TOMS, B.A.Sc., 60 Spencer Ave., Toronto, Ont.

1. H. W. TYE, Winnipeg, Man.
Construction Dept., C.P.R.

3. C. P. VAN NORMAN, Toronto, Ont.
Post-graduare Course in Engineering, University of Toronto.

1. T. L. VILLENEUVE, Chicoutimi, Que.
Assistant Engineer, Dept., of Public Works.

1. J. A. WALKER, B.A.Sc., Nelson, B.C.
With A. L. McCulloch, C.E., B.C.L.S.

3.*B. W. WAUGH, Toronto, Ont.
Post-graduate Course in Engineering, University of Toronto.

3. R. M. WEDLAKE, B.A.Sc., Brantford, Ont.
With Cockshutt Plow Co., Ltd.

3. R. P. WEIR, Toronto, Ont.
With C. H. Mitchell. Consulting Engineer.

1. A. M. WEST, B.A.Sc., Vancouver, B.C.
C.N.R. Office.

1. W. R. WHITE, Drayton, Ont.

3. W. J. WHITE, B.A.Sc., Boston, Mass.
Supt., on Construction, General Electric Co.

---

*Diploma with honours.

1908—*Continued.*

3.*F. D. WILSON,   Toronto, Ont.
    Draftsman, McGregor & McIntyre, Ltd.
1. J. M. WILSON,   Moose Jaw, Sask.
    City Engineer.
1. D. O. WING,   Prince Rupert, B.C.
    With G.T.P. Co.
3*R. YOUNG.   Lake Burnstzean, Burrard Inlet., B.C.

1909.

3. E. G. ARENS,   Niagara Falls, Ont.
    Draftsman with H. D. Symmes, Contractor.
3. H. V. ARMSTRONG,   Dunnville, Ont.
    Engineering Staff of Willis Chipman, C.E.
2.*E. T. AUSTIN,   Toronto. Ont.
    Post-graduate Course in Engineering, University of Toronto.
3. W. H. BARRY, B.A.Sc.,   Niagara Falls, Ont.
3. R. D. S. BECKSTEDT, B.A.Sc.,   Wilkinsburg, Pa.
3. R. E. BEITH,   Cochrane, Ont.
    With Transcontinental Ry.
1.*G. A. BENNETT, B.A.Sc.,   Box 1928, Calgary. Alta.
    Dominion Land Surveyor, Dept. of the Interior.
3. F. R. BIRCHARD, B.A.Sc ,   Barrie, Ont
    Canada Producer & Gas Engine Co.
3.W . D. BLACK, B.A.Sc.,   Montreal, Que.
    Supt., Otis-Fensom Elevator Co., Ltd.
3.*D. C. BLIZARD,   Toronto, Ont.
    Post-graduate Course in Engineering. University of Toronto.
1.*W. J. BOULTON,   Wallaceburg, Ont.
3. G. H. BOWEN, B.A.Sc.,   Niagara Falls, Ont.
    Engineer on Construction, Queen Victoria Park.
3. C. E. BROWN, B.A.Sc.,   Hamilton, Ont.
    Canadian Westinghouse Co.
1. E. W. BROWNE, B.A.Sc.,   247 Cannon St. E., Hamilton, Ont.
1. J. A. BUCHANAN,   Comber, Ont.
3. J. E. BURNS, B.A.Sc.,   231 Seaton St., Toronto, Ont.
1. M. G. CAMERON, B.A.Sc.,   Peterboro', Ont.
3.*R. A. CAMPBELL,   Alliston, Ont.
1. V. S. CHESNUT,   Mount Lebanon, B.C.
    Canadian Northern Ontario Ry.
1.*C. G. CLINE,   Toronto, Ont.
    Post-graduate Course in Engineering, University of Toronto.
1. J. G. COLLINSON. B.A.Sc.,   St. Thomas, Ont.
1. G. W. COLTHAM, B.A.Sc.,   Aurora, Ont.
3.*H. A. COOCH, B.A.Sc.,   Toronto, Ont.
    Demonstrator in Electrical Engineering, University of Toronto.

*Diploma with honours.

1909—*Continued.*

3. W. E. CORMAN, Stony Creek, Ont.
3. T. H. CROSBY, Toronto, Ont.
   Post-graduate Course in Engineering, University of Toronto.
3. R. H. CUNNINGHAM, Chicago, Ill.
   With Bryan-Marsh Co.
1.* F. A. DALLYN, B.A.Sc 1488 King St. W., Toronto, Ont.
3. C. N. DANKS, Sherbrooke, Que.
   Draftsman, Canadian Rand Co., Ltd.
1. E. M. DANN, Ottawa, Ont.
   Railway Lands Branch, Dept. of the Interior.
3. H. W. DAVIS, Kingston, Ont.
   With A. Davis & Son, Ltd., Leather Manufacturers.
2.*A. I. DAVIS, B.A.Sc., 82 Kendal Ave., Toronto, Ont.
1. H. C. DAVIS Burlington, Ont.
1. I. H. DAWSON, St. Catharines, Ont.
   Draftsman.
3. W. H. DELAHAYE, B. A. Sc., Pembroke, Ont
3. W. P. DERHAM, B.A.Sc., Ottawa, Ont.
5.*W. A. DODDS, B.A,S.c., Syracuse, N.Y.
   With Penman Littlehales Chemical Co.
1. R. H. DOUGLAS, Edmonton, Alta.
   Department of Public Works.
1. F. S. FALCONER, B.A.Sc., Shelburne, Ont.
3. T. A. FARGEY, B.A.Sc., Newark, N. J.
   With General Electric Co.
1. J. B. FERGUSON, Chicago, Ill.
3. A. T. FERGUSSON, B.A.Sc., Toronto, Ont.
3. T. E. FREEMAN, B.A.Sc., Peterboro', Ont.
   Canadian General Electric Co.
3. E. R. FROST, B.A.Sc., Toronto, Ont.
   With Smith, Keery & Chace.
1. A. E. GLOVER, Toronto, Ont.
   Post-graduate Coures in Engineering, University of Toronto.
5. A. E. GOODERHAM, Toronto, Ont.
1. D. A. GRAHAM, B.A.Sc., Ivan, Ont.
2. R. R. GRANT, 106 Warren Rd., Toronto, Ont.
1. J. E. GRAY, Toronto, Ont.
   Post-graduate Course in Engineering, University of Toronto.
1. G. E. D. GREENE, B.A.Sc., Toronto, Ont.
1. W. H. GREENE. 26 South Drive, Toronto, Ont.
1. W. W. GUNN, B.A.Sc., Montreal, Que.
   With Dominion Bridge Co.
3. C. J. HARPER, Pittsfield, Mass.
1. D. W. HARVEY, B.S.Ac., London, Ont.
1. C. O. HAY, 114 Liberty St., New York, N.Y
   Supt. on Construction, Muralt & Co.

*Diploma with honours.

1909—*Continued.*

3.*J. HEMPHILL,   Canton, Ohio.
    Canton Electric Power Co.

1.*G. HOGARTH,   Toronto, Ont.
    Engineer's Office, Dept. of Public Works of Ontario.

3. A. E. HOLMES, B.A.Sc.,   Hamilton, Ont.
    Canadian Westinghouse Co.

3. C. R. HOLMES,   Toronto, Ont.
    Post-graduate Course in Engineering, University of Toronto.

1. G. C. HOSHAL, B.A.Sc.,   Toronto, Ont.

3. C. HUGHES, B.A.Sc.,   Toronto, Ont.

1. A. E. HUNTER, B.A.Sc.,   Toronto, Ont.

3.*H. IRWIN, B.A.Sc.,   Toronto, Ont.
    Secretary, Engineering Society, University of Toronto.

3. J. ISBISTER,   B.A.Sc.,   Wingham, Ont.

3. F. P. JACKES, B.A.Sc.,   Toronto, Ont.
    With Canadian Westinghouse Co.

1.*J. E. JACKSON,   Oxford Centre, Ont.

1. E. W. JAMES, B.A.Sc.,   Toronto, Ont.

1.*C. C. JOHNSON, B.A.Sc.,   Toronto, Ont.
    City Hall.

1. C. E. JOHNSTON,   Toronto, Ont.
    Post-graduate Course in Engineering, University of Toronto.

1. W. J. JOHNSTON,   Porcupine, Ont.

1.*A. H. E. KEFFER,   North Bay, Ont.
    With T. & N.O. Ry.

3. J. B. O. KEMP, B.A.Sc.,   Toronto, Ont.
    Fellow in Drawing, University of Toronto.

3. W. R. KEY, B.A.Sc.,   Toronto, Ont.
    Mechanical Draftsman.

5. H. N. KLOTZ, B.A.Sc.,   Toronto, Ont.
    Chemist, Gutta Percha & Rubber Co.

3. A. W. LAMONT, B.A.Sc.,   Winnipeg, Man.
    Sales Engineer, Canadian Westinghouse Co., Ltd.

3.*C. B. LANGMUIR, B.A.Sc.,   Toronto, Ont
    Sales Dept., Factory Products, Ltd.

3. A. E. LENNOX, B.A.Sc.,   Cleveland, Ohio.
    National Electric Lamp Association.

1.*R. W. E. LOUCKS,   Delisle, Sask.
    Asst. to A. Fawcett, D.L.S.

1. N. C. A. LLOYD,

3. E. D. MACFARLANE, B.A.Sc.,   Cleveland, Ohio.

1. J. G. MACKINNON,   Toronto, Ont.
    Optical Dept., T. Eaton Co.

1. W. A. MACLACHLAN   Guelph, Ont.

3. B. A. MACLEAN, B.A.Sc.,   Hamilton, Ont.

1. N. W. MACPHERSON, B.A.Sc.,   St. Thomas, Ont.

   *Diploma with honours.

1909—*Continued.*

3. D. D. McALPINE, Toronto, Ont.
Post-Graduate Course in Engineering, University of Toronto.

1. A. S. McARTHUR, B.A.Sc., Toronto, Ont.

3. C. R. McCOLLUM, B.A.Sc., Toronto, Ont.
With Canada Cycle & Motor Co.

3.*A. S. McCORDICK, Toronto, Ont·
Post-Graduate Course in Engineering, University of Toronto.

3. P. J. McCUAIG, B.A.Sc., Milwaukee, Wis.

1. F. H. McKECHNIE, Cochrane, Ont.
Resident Engineer, Trans. Ry.

3. W. G. McINTOSH,.

3. G. McLEOD, Waupaca, Wis.
Electrician, Electric Light & Ry. Co.

1. V. McMILLAN, B.A.Sc., Toronto, Ont.

3. N. H. MANNING, B.A.Sc., Toronto, Ont.
Demonstrator in Mechanical Engineering. University of Toronto.

1.*A. B. MANSON, B.A.Sc., Stratford. Ont
City Engineer's Staff.

1.*E. S. MARTINDALE, Toronto, Ont.
Post-Graduate Course in Engineering, University of Toronto.

1. O. W. MARTYN, Toronto, Ont.
Post-Graduate Course in Engineering, University of Toronto.

2. C. A. MORRIS, B.A.Sc., Toronto, Ont.

3. G. MORTON, B.A.Sc.. Calgary, Alta.
Canadian Westinghouse Co.

1.*F. V. MUNRO, B.A.Sc., Toronto, Ont.
Fellow in Drawing, University of Toronto.

1. E. A. NEVILLE, Toronto, Ont.
Post-Graduate Course in Engineering, University of Toronto.

1. J. NEWTON, B.A.Sc., Toronto, Ont.
City Engineer's Department.

3.*L. S. ODELL, Toronto, Ont.
Fellow in Drawing, University of Toronto.

3. V. J. O'DONNELL, Toronto, Ont.

3. J. J. O'HEARN, Peterboro', Ont.
Canadian General Electric Co.

1. A. W. PAE, Calgary, Alta.
District Hydrographer.

1.*A. M: PETRY, B.A.Sc., Toronto, Ont.
Assistant Manager, " Chas. Potter."

1. R. B. PIGOTT, Toronto, Ont.
Post-Graduate Course in Engineering, University of Toronto.

2. G. M. PONTON, Blairmore, Alta.
With West Canadian Collieries Co.

3.*C. J. PORTER, B.A.Sc., Portland, Oregon.
Draftsman, Mount Hood Ry. & Power Co.

3. A. I. PROCTOR, Hamilton, Ont.

---

*Diploma with honours.

1909—*Continued.*

1. J. QUAIL,   Toronto, Ont.
1. A. F. RAMSPERGER,   Toronto, Ont.
   Draftsman, With Toronto Iron Works Ltd.
1.*C. R. REDFERN, B.A.Sc.,   Toronto, Ont.
   Demonstrator in Applied Mechanics. University of Toronto.
3.*L. T. RUTLEDGE, B.A.Sc.,   Toronto, Ont.
   Fellow in Drawing, University of Toronto.
1. A. U. SANDERSON, B.A.Sc.,   Toronto, Ont.
3.*R. A. SARA,   Toronto, Ont.
   Post-Graduate Course in Engineering, Univrsity of Toronto.
3.*A. SCHLARBAUM,   Toronto, Ont.
   Post-Graduate Course in Engineering, University of Toronto.
3.*C. SCHWENGER, B.A.Sc.,   Toronto, Ont.
   Electrical Department, City Hall.
1. C. A. SCOTT.   Toronto, Ont.
   Roadway Department, City Hall.
1. A. SEDGWICK,   Kaministikuia, Ont.
   Engineer, Dog Lake Storage Works.
1. B. H. SEGRE,   Winnipeg, Man.
1. F. V. SEIBERT,   Toronto, Ont.
   Fellow in Surveying, University of Toronto.
5. M. R. SHAW,   Midland, Mich.
   Chemist, Dow Chemical Co.
3. M. W. SPARLING,   Toronto, Ont.
   Post-Graduate Course in Engineering, University of Toronto.
3. J. J. SPENCE,   Toronto, Ont.
1. D. S. STAYNER, B.A.Sc.,   Toronto, Ont.
   Bridge Department, City Engineer's Office.
1.*N. C. STEWART,   Toronto, Ont.
   Post-Graduate Course in Engineering, University of Toronto.
1.*P. H. STOCK,   St. Catharines, Ont.
   Resident Engineer, N. St. C. & T. R. Ry.
1. J. C. STREET,   Toronto, Ont.
   Post-Graduate Course in Engineering, University of Toronto.
3. S. STROUD, B.A.Sc.,   East Pittsburgh, Pa.
   Westinghouse Electric & Manufacturing Co.
1. C. C. SUTHERLAND,   Toronto, Ont.
   Post-Graduate Course in Engineering, Unwersity of Toronto.
1. R. G. SWAN, B.A.Sc.,   Toronto, Ont.
1. A. D. SWORD,   Toronto, Ont.
   Post-Graduate Course in Engineering, University of Toronto.
1.*H. W. TATE, B.A.Sc.,   Toronto, Ont.
3.*E. A. THOMPSON,   Toronto, Ont.
   With Smith, Kerry & Chace.
1. G. A. TIPPER, B.A.Sc.,   Toronto, Ont.
3. A. G. TREES, B.A.Sc.,   Toronto, Ont.

*Diploma with honours.

1909—*Continued.*

3. W. G. TURNBULL, B.A.Sc.,   Milwaukee, Wis.
  The Cutler Hammer Mfg. Co.
1. J. E. UNDERWOOD,   Lakelet, Ont.
1. C. P. VANNORMAN.   Toronto, Ont.
  Post-Graduate Course in Engineering, University of Toronto.
1. J. VANNOSTRAND,   Toronto, Ont.
  Student.
1. A. VATCHER, B.A.Sc.,   Freshwater, Bay deVerde. Nfl.
  With the Reid Newfoundland Co.
1. C. M. WALKER,   Toronto, Ont.
  Post-Graduate Course in Engineering, University of Toronto.
1. E. E. WEBB,   Orillia, Ont.
1. C. E. WEBB, B.A.Sc.,   Toronto, Ont.
3. F. C. WHITE,   Toronto, Ont.
  Post-Graduate Course in Engineering, University of Toronto.
3. A. R. WHITELAW,   Toronto, Ont.
  With Smith, Kerry & Chace.
1. R. G. WILKINSON,   Aberarder, Ont.
5.*J. A. McK. WILLIAMS, B.A.Sc.,   Toronto, Ont.
1.*O. T. G. WILLIAMSON, B.A.Sc.,   Guelph, Ont.
3. L. R. WILSON.   Toronto, Ont.
  Post-Graduate Course in Engineering, University of Toronto.
3. F. F. WILSON,   Toronto, Ont.
  Post-Graduate Course in Engineering, University of Toronto.
2. S. A. WOOKEY,   Toronto, Ont.
  Post-Graduate Course in Engineering, University of Toronto.

1910.

2. J. H. ADAMS,   Toronto, Ont.
  Post-Graduate Course in Engineering, University of Toronto.
3.*O. F. ADAMS,   Toronto, Ont.
  Fellow in Electrical Engineering, University of Toronto.
1.*W. G. AMSDEN,   Toronto, Ont.
  Post-Graduate Course in Engineering, University of Toronto.
1. J. A. BAIRD,   Toronto, Ont.
  Post-Graduate Course in Engineering, University of Toronto.
1.*W. J. BAIRD.
1. H. A. BARNETT,   Toronto, Ont.
  Post-Graduate Course in Engineering, University of Toronto.
1.*E. W. BERRY,   Seaforth, Ont.
1.*H. C. BINGHAM,   Moose Jaw Sask.
  City Engineer's Department.
2. D. G. BISSET,   Toronto, Ont.
  Post-Graduate Course in Engineering, Univers-ty of Toronto.
1. R. H. H. BLACKWELL,   Toronto, Ont.
1.*E. P. BOWMAN,   Toronto, Ont.
  Post-Graduate Course in Engineering, University of Toronto.

---

*Diploma with honours.

1910—*Continued.*

2. A. F. Brock,   Toronto, Ont.
   Post-Graduate Course in Engineering, University of Toronto.
3. M. O. Browne,   Toronto, Ont.
3. J. R. Burgess,   Toronto, Ont.
   Post-Graduate Course in Engineering, University of Toronto.
1. N. G. H. Burnham,   Toronto, Ont.
3.*W. C. Cale,   Glenwood Springs Colo.
   With Central Colorado Power Co.
2.*A. D. Campbell,   Toronto, Ont.
   Post-Graduate Course in Engineering, University of Toronto.
3. W. M. Carlyle,   Toronto, Ont.
   Post-Graduate Course in Engineering, University of Toronto.
3. N. S. Caudwell,   Toronto, Ont.
1.*D. C. Chisholm,   Toronto, Ont.
   Post-Graduate Course in Engineering, University of Toronto.
1. J. A. Claveau,   Chicoutimi, P.Q.
3. L. S. Cockburn,   Toronto, Ont.
   Post-Graduate Course in Engineering. University of Toronto.
3. A. G. Code,   Toronto, Ont.
   Post-Graduate Course in Engineering, University of Toronto.
3. C. R. Cole,   Woodstock, Ont.
1. G. A. Colquhoun,   Toronto, Ont.
   Post-Graduate Course in Engineering, University of Toronto.
4.*J. H. Craig,   Toronto, Ont.
   Fellow in Drawing, University of Toronto.
3.*C. D. Dean,   Toronto, Ont·
   Post-Graduate Course in E gineering, University of Toronto.
3. R. L. Dobbin,   Toronto, Ont.
   Post-Graduate Course in Engineering, University of Toronto.
3.*W. P. Dobson,   Toronto, Ont.
   Post-Graduate Course in Engineering, University of Toronto.
3.*J. M. Duncan,   Toronto, Ont.
2. V. H. Emery,   Toronto, Ont.
   Post-Graduate Course in Engineering, University of Toronto.
3. W. J. Evans,   Jermyn, Ont.
3. H. W. Fairlie,   Toronto, Ont.
   Tungsiolier Co., of Canada.
3.*C. R. Ferguson,   Toronto, Ont.
   Post-Graduate Course in Engineering, University of Toronto.
3. J. W. Ferguson,   Toronto, Ont.
   Post-Graduate Course in Engineering, University of Toronto.
4.*J. B. K. Fisken,   Toronto, Ont.
   Post-Graduate Course in Engineering, University of Toronto.
1. A. W. Fletcher,   Toronto, Ont.
   Post-Graduate Course in Engineering, University of Toronto.
1.*J. A. Fletcher,   Fisher River, Man.
   Assistant to E. W. Robinson, D.L.S.

*Diploma with honours.

1910—*Continued.*

3. F. T. FLETCHER, Toronto, Ont.
Post-Graduate Course in Engineering, University of Toronto.
3. T. R. C. FLINT,.
3. R. C. FOLLETT,
2. J. M. FOREMAN, Toronto, Ont.
Post-Graduate Course in Engineering, University of Toronto.
1. W. J. FOSTER,
3.*W. C. FOULDS, Toronto, Ont.
Post-Graduate Course in Engineering, University of Toronto.
1. A. FRASER, Toronto, Ont.
. *Post-Graduate Course in Engineering, University of Toronto.*
2. J. FREDIN, London, Ont.
1. M. M. GIBSON, Toronto, Ont.
Post-Graduate Course in Engineering, University of Toronto.
1. V. A. E. GOAD, Toronto, Ont·
Post-Graduate Course in Engineering, University of Toronto.
3. V. S. GOODEVE, Toronto, Ont
2. W. A. GORDON, Sundridge, Ont.
3. V. F. GOURLAY, Toronto, Ont.
Post-Graduate Course in Engineering, University of Toronto.
2. R. L. GREENE. Toronto, Ont.
Post-Graduate Course in Engineering, University of Toronto.
5. J. H. HARRIS, Toronto, Ont.
Post-Graduate Course in Engineering. University of Toronto.
1.*N. J. HARVIE, Toronto, Ont.
Post-Graduate Course in Engineering, University of Toronto.
1. J. G. HELLIWELL, Toronto, Ont.
1. J. F. HENDERSON, Toronto, Ont.
3. F. G. HICKLING, Wilkinsburg, Pa.
2.*P. E. HOPKINS, Toronto, Ont.
Post-Graduate Course in Engineering, University of Toronto.
3.*W. J. IRWIN, East Pittsburg, Pa.
Apprenticeship Course, Westinghouse Machine Co.
2. F. L. JAMES, Toronto, Ont.
Post-Graduate Course in Engineering. University of Toronto.
1. H. C. JOHNSTON,. Toronto, Ont.
Post-Graduate Course in Engineering, University of Toronto.
1. R. H. JOHNSON, Toronto, Ont.
Post-Graduate Course in Engineering, University of Toronto.
1. J. C. KEITH, Toronto, Omt.
Post-Graduate Course in Engineering, University of Toronto.
2.*J. T. KING, Toronto, Ont.
Fellow in Mining Engineering, University of Toronto.
3. G. A. KINGSTONE, Toronto, Ont.
Post-Graduate Course in Engineering, University of Toronto.
2. G. L. KIRWIN, Toronto, Ont.
Post-Graduate Course in Engineering. University of Toronto.

---
*Diploma with honours.

1910—*Continued.*

5. P. T. Kirwin,	Toronto, Ont.
Fellow in Chemistry, University of Toronto.

1. S. Knight,	Toronto, Ont.
Post-Graduate Course in Engineering, University of Toronto.

3. E. R. Lawler,	Toronto, Ont.

3.*C. B. Leaver,	Toronto, Ont.
Post-Graduate Course in Engineering, University of Toronto.

3. R. G. Lee,	Toronto, Ont.
Post-Graduate Course in Engineering, University of Toronto.

1. J. C. Longstaff,	Toronto, Ont.

3. J. B. MacDonald,	Toronto, Ont.
Post-Graduate Course in Engineering, University of Toronto.

2.*A. D. Macdonald,

1. J. A. MacDonald,	Toronto, Ont.
Post-Graduate Course in Engineering, University of Toronto.

1. G. A. MacDonald,	Toronto, Ont.
Post-Graduate Course in Engineering, University of Toronto.

1.*A. E. MacGregor,	Wilkeson, Wash.
Analyst & Draftsman, Wilkeson Coal & Coke Co.

1. E. G. MacKay,	Hamilton, Ont.

1.*G. G. MacLennan,	Prince Albert ,Sask.
Assistant to A. St. Cyr, D.L.S.

1. D. D. MacLeod,	Toronto, Ont.
Post-Graduate Course in Engineering, University of Toronto.

3. H. G. MacMurchy,	Toronto, Ont.
Post-Graduate Course in Ebgineering, Unwersity of Toronto.

3.*H. J. MacTavish,	Toronto, Ont.
Post-Graduate Course in Engineering, University of Toronto.

4. T. C. McBride,	Toronto, Ont.
Post-Graduate Course in Engineering, University of Toronto.

1. S. G. McDougall,	Toronto, Ont.
Post-Graduate Course in Engineering, University of Toronto.

1.*T. A. McElhanney,	Kincardine, Ont.

1.*P. J. McGarry,	Merriton, Ont.

3.*L. R. McKim,	Wyecombe, Ont.

1.*J. McNiven,	Toronto, Ont.
Post-Graduate Course in Engineering, University of Toronto.

3. J. I. McSloy,	Toronto, Ont.
Post-Graduate Course in Engineering. University of Toronto.

2. A. W. R. Maisonville.	Toronto, Ont.
Post-Graduate Course in Engineering, University of Toronto.

1.*N. Marr,	Cochrane, Ont.
Engineering Staff, Transcontinental Ry.

1.*W. H. Martin,	Toronto, Ont.
Post-Graduate Course in Engineering, University of Toronto.

2. A. C. Matthews,	Toronto, Ont.
Post-Graduate Course in Engineering, University of Toronto.

*Diploma with honours.

1910—*Continued.*

3. *H. O. MERRIMAN,   Toronto, Ont.
Post-Graduate Course in Engineering, University of Toronto.

1. *D. J. MILLER,   Orillia, Ont.

1. F. S. MILLIGAN,   Toronto, Ont.
Post-Graduate Course in Engineering, University of Toronto.

3. P. E. MILLS,   Toronto, Ont.
Post-Graduate Course in Engineering, University of Toronto.

3. J. P. MORGAN,   Newmarket, Ont.

1. A. H. MUNRO,   Toronto, Ont.
. Post-Graduate Course in Engineering, University of Toronto.

3. J. C. NASH,   Hamilton, Ont.
Draftsman, Canadian Westinghouse Co.

1. *V. A. NEWHALL,   Toronto, Ont.
Post-Graduate Course in Engineering, University of Toronto.

2. *W. E. NEWTON,   Toronto, Ont.
Post-Graduate Course in Engineering, University of Toronto.

1. F. T. NICHOL,   Toronto, Ont.
Post-Graduate Course in Engineering, University of Toronto.

1. C. M. O'NEIL,   Toronto, Ont.
Post-Graduate Course in Engineering, University of Toronto.

3. C. E. PALMER,   Toronto, Ont.
Post-Graduate Course in Engineering, University of Toronto.

3. G. C. PARKER,   Toronto, Ont.
Post-Graduate Course in Engineering, University of Toronto.

3. K. K. PEARCE,   Toronto, Ont.
Post-Graduate Course in Engineering, University of Toronto.

3. C. H. PHILLIPS,   Toronto, Ont.
Post-Graduate Course in Engineering, University of Toronto.

1. D. E. PYE,   Cranbrook, B.C.

1. W. S. RAMSAY,   Toronto, Ont.
Post-Graduate Course in Engineering, University of Toronto

3. B. J. REDFERN,   Toronto, Ont·

1. H. C. RITCHIE,   Calgary, Alta.

1. O. W. ROSS,   Lachine, P.Q.
With Dominion Bridge Co.

1. W. F. B. RUBIDGE,   Dixie, Ont.

3. W. C. SHAW,   Toronto, Ont.

1. *W. C. SMITH,   Two Harbors, Minn.
With D. & I. R.R. Co.

5. G. E. SMITH,   Toronto, Ont.

2. R. J. SPRY,   Greenwood, B.C·
Metallurgist, The B. C. Copper Co.

2. A. L. STEELE,   Toronto, Ont.
Post-Graduate Course in Engineering, University of Toronto.

2. *H. M. STEVEN,   Toronto, Ont·
Post-Graduate Course in Engineering. University of Toronto.

1. *L. I. STONE,   Toronto, Ont.

---

*Diploma with honours.

1910—*Continued.*

3. A. L. SUTHERLAND,　　　　　　　　　　　　　Toronto, Ont.
　　Post-Graduate Course in Engineering, University of Toronto.
3. E. A. TERNAN,　　　　　　　　　　　　　　Toronto, Ont.
　　Engineering Dept., Canadian General Electric Co.
5.*W. H. THOM,　　　　　　　　　　　　　　Watford, Ont.
3. H. B. THOMPSON,　　　　　　　　　　　Wellington, Ont.
3. R. M. A. THOMPSON,　　　　　　　　　　Toronto, Ont.
　　Post-Graduate Course in Engineering, University of Toronto.
2.*C. G. TITUS,　　　　　　　　　　　　Gowganda, Ont.
　　Manager, The Bartlett Mines.
3. K. M. VANALLEN,　　　　　　　　　　　Toronto, Ont.
　　Post-Graduate Course in Engineering, University of Toronto.
1. L. T. VENNEY,　　　　　　　　　　　　Toronto, Ont.
　　Post-Graduate Course in Engineering, University of Toronto.
1. N. WAGNER,　　　　　　　　　　　　Toronto, Ont.
　　Bridge Dept., Canada Foundry Co.
1. R. M. WALKER.
2. T. WALTON,　　　　　　　　　　　　Toronto, Ont.
　　Post-Graduate Course in Engineering, University of Toronto.
1. G. A. WARRINGTON,　　　　　　　　　Toronto, Ont.
　　Post-Graduate Course in Engineering, University of Toronto.
3. M. B. WATSON,　　　　　　　　　　Toronto. Ont.
　　Post-Graduate Course in Engineering, University of Toronto.
3.*H. M. WHITE,　　　　　　　　　　　Chatham, Ont.
4. W. S. WICKENS,　　　　　　　　　　Toronto, Ont.
　　Post-Graduate Course in Engineering, University of Toronto.
3.*G. K. WILLIAMS,　　　　　　　　　　Toronto, Ont.
　　Post-Graduate Course in Engineering, University of Toronto.
1.*W. H. WILSON,　　　　　　　　　　Toronto, Ont.
　　Fellow in Physics, University of Toronto.
1. G. R. WORKMAN,　　　　　　　　　Grand Mere, Que.
　　With the Laurentide Paper Co., Ltd.
3. L. A. WRIGHT,　　　　　　　　　　Toronto, Ont.
　　Post-Graduate Course in Engineering, University of Toronto.
3.*A. W. YOUELL,　　　　　　　　　　Toronto, Ont.
　　Post-Graduate Course in Engineering, University of Toronto.
1. W. S. YOUNG,　　　　　　　　　　Toronto, Ont.
　　Post-Graduate Course in Engineering, University of Toronto.

*Diploma with honours.

26 22

5 8

Lightning Source UK Ltd.
Milton Keynes UK
UKHW010849111218
333785UK00016B/1552/P